Reviews of
Environmental Contamination
and Toxicology

VOLUME 225

Reviews of Environmental Contamination and Toxicology

Editor
David M. Whitacre

VOLUME 225

Coordinating Board of Editors

ISSN 0179-5953
ISBN 978-1-4899-8820-1   ISBN 978-1-4614-6470-9 (eBook)
DOI 10.1007/978-1-4614-6470-9
Springer New York Heidelberg Dordrecht London
© Springer Science+Business Media New York 2013
Softcover re-print of the Hardcover 1st edition 2013

Printed on acid-free paper

Springer is part of Springer Science+Business Media (www.springer.com)

Foreword

International concern in scientific, industrial, and governmental communities over traces of xenobiotics in foods and in both abiotic and biotic environments has justified the present triumvirate of specialized publications in this field: comprehensive reviews, rapidly published research papers and progress reports, and archival documentations. These three international publications are integrated and scheduled to provide the coherency essential for nonduplicative and current progress in a field as dynamic and complex as environmental contamination and toxicology. This series is reserved exclusively for the diversified literature on "toxic" chemicals in our food, our feeds, our homes, recreational and working surroundings, our domestic animals, our wildlife, and ourselves. Tremendous efforts worldwide have been mobilized to evaluate the nature, presence, magnitude, fate, and toxicology of the chemicals loosed upon the Earth. Among the sequelae of this broad new emphasis is an undeniable need for an articulated set of authoritative publications, where one can find the latest important world literature produced by these emerging areas of science together with documentation of pertinent ancillary legislation.

Research directors and legislative or administrative advisers do not have the time to scan the escalating number of technical publications that may contain articles important to current responsibility. Rather, these individuals need the background provided by detailed reviews and the assurance that the latest information is made available to them, all with minimal literature searching. Similarly, the scientist assigned or attracted to a new problem is required to glean all literature pertinent to the task, to publish new developments or important new experimental details quickly, to inform others of findings that might alter their own efforts, and eventually to publish all his/her supporting data and conclusions for archival purposes.

In the fields of environmental contamination and toxicology, the sum of these concerns and responsibilities is decisively addressed by the uniform, encompassing, and timely publication format of the Springer triumvirate:

Reviews of Environmental Contamination and Toxicology [Vol. 1 through 97 (1962–1986) as Residue Reviews] for detailed review articles concerned with any aspects of chemical contaminants, including pesticides, in the total environment with toxicological considerations and consequences.

Bulletin of Environmental Contamination and Toxicology (Vol. 1 in 1966) for rapid publication of short reports of significant advances and discoveries in the fields of air, soil, water, and food contamination and pollution as well as methodology and other disciplines concerned with the introduction, presence, and effects of toxicants in the total environment.

Archives of Environmental Contamination and Toxicology (Vol. 1 in 1973) for important complete articles emphasizing and describing original experimental or theoretical research work pertaining to the scientific aspects of chemical contaminants in the environment.

Manuscripts for Reviews and the Archives are in identical formats and are peer reviewed by scientists in the field for adequacy and value; manuscripts for the *Bulletin* are also reviewed, but are published by photo-offset from camera-ready copy to provide the latest results with minimum delay. The individual editors of these three publications comprise the joint Coordinating Board of Editors with referral within the board of manuscripts submitted to one publication but deemed by major emphasis or length more suitable for one of the others.

Coordinating Board of Editors

Preface

The role of *Reviews* is to publish detailed scientific review articles on all aspects ofenvironmental contamination and associated toxicological consequences. Such articlesfacilitate the often complex task of accessing and interpreting cogent scientificdata within the confines of one or more closely related research fields.

In the nearly 50 years since *Reviews of Environmental Contamination andToxicology* (formerly *Residue Reviews*) was first published, the number, scope, andcomplexity of environmental pollution incidents have grown unabated. During thisentire period, the emphasis has been on publishing articles that address the presenceand toxicity of environmental contaminants. New research is published each yearon a myriad of environmental pollution issues facing people worldwide. This fact,and the routine discovery and reporting of new environmental contamination cases,creates an increasingly important function for *Reviews*.

The staggering volume of scientific literature demands remedy by which data canbe synthesized and made available to readers in an abridged form. *Reviews* addressesthis need and provides detailed reviews worldwide to key scientists and science orpolicy administrators, whether employed by government, universities, or the privatesector.

There is a panoply of environmental issues and concerns on which many scientistshave focused their research in past years. The scope of this list is quitebroad, encompassing environmental events globally that affect marine and terrestrialecosystems; biotic and abiotic environments; impacts on plants, humans, andwildlife; and pollutants, both chemical and radioactive; as well as the ravages ofenvironmental disease in virtually all environmental media (soil, water, air). Newor enhanced safety and environmental concerns have emerged in the last decade tobe added to incidents covered by the media, studied by scientists, and addressedby governmental and private institutions. Among these are events so striking thatthey are creating a paradigm shift. Two in particular are at the center of everincreasingmedia as well as scientific attention: bioterrorism and global warming.Unfortunately, these very worrisome issues are now superimposed on the alreadyextensive list of ongoing environmental challenges.

The ultimate role of publishing scientific research is to enhance understandingof the environment in ways that allow the public to be better informed. Theterm "informed public" as used by Thomas Jefferson in the age of enlightenmentconveyed the thought of soundness and good judgment. In the modern sense, being"well informed" has the narrower meaning of having access to sufficient information. Because the public still gets most of its information on science and technologyfrom TV news and reports, the role for scientists as interpreters and brokers of scientificinformation to the public will grow rather than diminish. Environmentalismis the newest global political force, resulting in the emergence of multinational consortiato control pollution and the evolution of the environmental ethic.Will the newpolitics of the twenty-first century involve a consortium of technologists and environmentalists,or a progressive confrontation? These matters are of genuine concernto governmental agencies and legislative bodies around the world.

For those who make the decisions about how our planet is managed, there is anongoing need for continual surveillance and intelligent controls to avoid endangeringthe environment, public health, and wildlife. Ensuring safety-in-use of the manychemicals involved in our highly industrialized culture is a dynamic challenge, forthe old, established materials are continually being displaced by newly developedmolecules more acceptable to federal and state regulatory agencies, public healthofficials, and environmentalists.

Reviews publishes synoptic articles designed to treat the presence, fate, and, ifpossible, the safety of xenobiotics in any segment of the environment. These reviewscan be either general or specific, but properly lie in the domains of analytical chemistryand its methodology, biochemistry, human and animal medicine, legislation,pharmacology, physiology, toxicology, and regulation. Certain affairs in food technologyconcerned specifically with pesticide and other food-additive problems mayalso be appropriate.

Because manuscripts are published in the order in which they are received infinal form, it may seem that some important aspects have been neglected at times. However, these apparent omissions are recognized, and pertinent manuscripts arelikely in preparation or planned. The field is so very large and the interests in itare so varied that the editor and the editorial board earnestly solicit authors andsuggestions of underrepresented topics to make this international book series yetmore useful and worthwhile.

Justification for the preparation of any review for this book series is that it dealswith some aspect of the many real problems arising from the presence of foreignchemicals in our surroundings. Thus, manuscripts may encompass case studies fromany country. Food additives, including pesticides, or their metabolites that may persistinto human food and animal feeds are within this scope. Additionally, chemicalcontamination in any manner of air, water, soil, or plant or animal life is within theseobjectives and their purview.

Manuscripts are often contributed by invitation. However, nominations for newtopics or topics in areas that are rapidly advancing are welcome. Preliminary communicationwith the editor is recommended before volunteered review manuscriptsare submitted.

Summerfield, North Carolina   David M. Whitacre

Contents

Microbial Transformation of Trace Elements in Soils in Relation to Bioavailability and Remediation

Nanthi S. Bolan, Girish Choppala, Anitha Kunhikrishnan,
Jinhee Park, and Ravi Naidu

Contents

N.S. Bolan (✉) • G. Choppala • R. Naidu
Centre for Environmental Risk Assessment and Remediation,
University of South Australia, Mawson Lakes, SA, Australia

Cooperative Research Centre for Contamination Assessment and Remediation
of the Environment, Adelaide, SA, Australia
e-mail: nanthi.bolan@unisa.edu.au

A. Kunhikrishnan
Chemical Safety Division, Department of Agro-Food Safety, National Academy
of Agricultural Science, Suwon-si, Gyeonggi-do, Republic of Korea

J. Park
Centre for Mined Land Rehabilitation, University of Queensland, St Lucia, QLD, Australia

D.M. Whitacre (ed.), *Reviews of Environmental Contamination and Toxicology,*
Reviews of Environmental Contamination and Toxicology 225,
DOI 10.1007/978-1-4614-6470-9_1, © Springer Science+Business Media New York 2013

1   Introduction

The term "trace elements" generally includes elements (both metals and metalloids) that occur in natural and perturbed environments in small amounts and that, when present in sufficient bioavailable concentrations, are toxic to living organisms (Adriano 2001). This group includes both biologically essential [e.g., cobalt (Co), copper (Cu), chromium (Cr), manganese (Mn), and zinc (Zn)] and nonessential [e.g., cadmium (Cd), lead (Pb), and mercury (Hg)] elements. The essential elements (for plant, animal, or human nutrition) are required in low concentrations and hence are known as "micro nutrients". The nonessential elements are phytotoxic and/or zootoxic and are widely known as "toxic elements" (Adriano 2001). Both groups are toxic to plants, animals, and/or humans at exorbitant concentrations (Alloway 1990; Adriano 2001). Heavy metal(loid)s, which include elements with an atomic density greater than 6 g cm^{-3} [with the exception of arsenic (As), boron (B), and selenium (Se)] are also considered to be trace elements.

Soil represents the major sink for trace elements released into the biosphere through both geogenic (i.e., weathering or pedogenic) and anthropogenic (i.e., human activities) processes. The mobility and bioavailability of trace elements in soils are affected by adsorption onto mineral surfaces, precipitation as salts, formation of stable complexes with organic compounds, and biotransformation by microorganisms (Alexander 2000; Adriano 2001). Soil is a biologically active integral component of the terrestrial ecosystem in which higher plants, soil constituents, and soil organisms interact, and the available energy in the form of organic and inorganic compounds promotes microbial activity and microbial weathering processes. The dynamics of trace elements in soils depend not only on their physicochemical interactions with inorganic and organic soil constituents but also on biological interactions largely associated with the microbial activities of soil–plant systems (Adriano 2001). Traditionally, most research efforts have focused on the physicochemical interactions of trace elements with soil components (Ross 1994; Dube et al. 2001; Pédrot et al. 2008). Only in recent times has the importance of microorganism–trace element interaction been realized, in relation to environmental health, ecotoxicology, and remediation (Adriano et al. 2004; Gadd 2010; Park et al. 2011a).

Two approaches have been used to examine the interaction between microbes and trace elements in soils (Alexander 1999): (1) the influence of trace elements on microbial populations and functions (e.g., biological nitrogen (N) fixation and nitrification reaction) and (2) the influence and role of microbes on the transformation of trace elements (e.g., redox reactions and bioaccumulation). Many researchers have examined the toxic effects of trace elements on microbial population and functions, the environmental factors affecting the toxicity, and the mechanisms involved in the development of trace-element resistance in microorganisms (Nies 1999; Smolders et al. 2004; Wang et al. 2007). Microorganisms control the transformation (microbial or biotransformation) of trace elements by various mechanisms that include oxidation, reduction, methylation, demethylation, complex formation, and biosorption (Alexander 1999).

Microbial transformation plays a key role in the behavior and fate of toxic trace elements, especially As, Cr, Hg, and Se in soils and sediments. Microbial transformation processes can influence the solubility and subsequent mobility of these trace elements in soils by altering their speciation and oxidation/reduction state (Gadd 2010). These processes play a major role in the bioavailability, mobility, ecotoxicology, and environmental health of these trace elements. For example, microbial reduction/methylation of trace elements and its consequences to human health has received attention primarily from a series of widespread poisoning incidents (Adriano 2001; Adriano et al. 2004). First, many cases of "Gasio-gas" poisoning resulted from converting arsenic trioxide (As_2O_3) in wallpaper glue into volatile poisonous trimethyl arsine or "Gasio-gas" [$(CH_3)_3As$]. Second, human poisoning at Minamata Bay and Niigata in Japan (Minamata disease), in the late 1950s, was believed to arise from the ingestion of fish and shellfish containing methylmercuric compounds that were derived from the biomethylation of mercuric salts by aquatic organisms. More recently, As contamination of surface- and ground-waters, mediated through redox reactions of geogenic As, became a major socio-political issue at several points around the globe (Mahimairaja et al. 2005).

Microbial reduction and methylation reactions have also been identified as important mechanisms for detoxifying toxic elements (Zhang and Frankenberger 2003). These processes are particularly important for those elements (e.g., As, Hg, and Se) that are able to form methyl or metal(loid)-hydride compounds. For example, microbial production (i.e., biogenic origin) of volatile compounds (e.g., alkyl selenides) is recognized as an important source of atmospheric trace element input, and it is estimated that approximately 15% of the total annual As and Se inputs to the atmosphere is derived from biogenic sources (Haygarth et al. 1994; Pécheyran et al. 1998; Kosolapov et al. 2004). Similarly, a biological conversion process (i.e., methylation) is the key to the biomagnification of Hg in the aquatic ecosystem (Ciesielski et al. 2010).

Thus, a greater understanding of microbial transformation processes will help to monitor the environmental fate of the trace elements, particularly through the food web, and will help to develop *in situ* bioremediation technologies that are environmentally compatible. It is unlikely that the natural phenomenon (e.g., natural attenuation) is optimal for the removal of toxic trace elements from contaminated sites. Microbial transformation processes can be readily managed and enhanced for efficient removal of contaminants, provided the biochemistry of these processes is understood. There is a growing interest in the potential exploitation of microorganisms in detoxifying contaminated sites, wastewater treatment, and metal(loid)s extraction from low-grade ores.

The aim of this review is to examine the influence of microbial processes on the accumulation and transformation of trace elements (with emphasis on As, Cr, Hg, and Se). We first briefly address the sources and speciation of these four major trace elements that are subject to soil microbial transformation. We then examine their microbial transformation in the context of the practical implications they have for remediation and ecotoxicology.

2   Sources and Speciation of Trace Elements in Soils and Sediments

The uses of As, Cr, Hg, and Se, and their species in soils and sediments are given in Table 1.

2.1   *Arsenic*

Arsenic is a naturally occurring element, the major source of which is weathering of igneous and sedimentary rocks, including coal. Coal is estimated to release 45,000 tonnes of As annually, while human activities release approximately 50,000 tonnes (Fergusson and Gavis 1972; Mahimairaja et al. 2005). Significant anthropogenic sources of As include fossil fuel combustion, leaching from mining wastes and landfills, mineral processing, and metal(loid) production. Precipitation from the atmosphere and application of a range of agricultural by-products (e.g., poultry manure) also contribute large quantities of As to the land (Christen 2001). Although the anthropogenic As source is increasingly becoming important, the recent episode of extensive As-contamination of groundwaters in Bangladesh and West Bengal is of geological origin, transported by rivers from sedimentary rocks in the Himalayas over tens of thousands of years (Mahimairaja et al. 2005).

In soils, As is present as arsenite [As(III)], arsenate [As(V)], and organic As (monomethyl arsenic acid and dimethyl arsenic acid or cacodylic acid) (Sadiq 1997; Smith et al. 1998; Mahimairaja et al. 2005). Arsenic species are adsorbed onto Fe, Mn, and Al compounds (Smith et al. 1998). Sorption of As species by organic matter and humic acid is also possible (Mahimairaja et al. 2005; Warwick et al. 2005). The main solid phase As species in soils and sediments include aluminum arsenate, iron arsenate, and calcium arsenate. In aquatic systems, As is predominantly bound to sediments. The arsenic concentration in suspended solids and sediments is many times higher than that in water, indicating that the suspended solids are a good scavenging agent and sediments a sink for As (Mahimairaja et al. 2005). Arsenic oxidation from As(III) to As(V) is a natural process that helps to alleviate toxicity in aquatic environments because As(V) is adsorbed onto the sediments and becomes relatively immobilized (Aposhian et al. 2003; Rubinos et al. 2011).

Table 1 Uses of arsenic, chromium, mercury, and selenium and their species in soils

Trace element	Uses	Species in soil and sediments	
		Acidic	Alkaline
Arsenic	Agriculture: Herbicides, insecticides, defoliants, wood preservatives, debarking tress and soil sterilant	As(V): $H_2AsO_4^-$; As(III): $As(OH)_3$	As(V): $HAsO_4^{2-}$; As(III): AsO_3^{3-}
	Livestock: Feed additives, disease prevention, cattle and sheep dips and algaecides		
	Medicine: Anti-syphilitic drugs, treatment of trypanosomiasis, amebiasis and sleeping sickness		
	Electronics: Solar cells, optoelectronic devices, semiconductors and light emitting diodes		
	Industry: Glassware, electrophotography, catalysts, pyrotechnics, antifouling paints, dyes and soaps, ceramics and pharmaceutical substances		
	Metallurgy: Alloys and battery plates		
Chromium	Livestock: Feed additives	Cr(VI): CrO_4^{2-}; Cr(III): $Cr(OH)^{2+}$	Cr(VI): CrO_4^{2-}; Cr(III): $Cr(OH)_4^-$
	Medicine: Cr(III) picolinate to treat diabetes, antidepressant and muscle growth		
	Industry: Wood tanalyzing; skin tanning, refractory bricks, audio cassettes, pigment, textile industries as mordants, anodizing aluminum, stainless steel, alloys, glass, printing inks, drilling muds, pyrotechnics, water treatment, cement producing plants and chemical synthesis		
	Medicine: Cr(III) picolinate to treat diabetes, antidepressant and muscle growth		
Mercury	Agriculture: seed fumigant and soil sterilant	Hg(II): Hg^{2+}; $Hg(Cl)_2$; CH_3Hg^+	Hg(II): $Hg(OH)_2$
	Medicine: Dental care, blood pressure cuffs, cosmetics (whitening additive)		
	Industry: gold extraction, chlorine (Cl_2) production, Portland cement, sulfuric acid, caustic soda, mercuric oxide batteries and lamps (fluorescent and high-intensity discharge lamps)		
Selenium	Agriculture: Fertilizer, constituent of fungicides	Se(VI): SeO_4^{2-}; Se(IV): $HSeO_3^-$	Se(VI): SeO_4^{2-}; Se(IV): SeO_3^{2-}
	Livestock: Feed additives for poultry and livestock		
	Industry: Manganese electrolysis, glass production and alloys		
	Electronics: solar cells, photocells, solar cells, Se rectifiers, LEDs, X-ray crystallography and toning of photographs		
	Medicine: Anti-dandruff shampoos and radioactive Se in diagnostic medicine		

2.2 Chromium

Chromium reaches the soil environment via industrial waste disposal from coal-fired power plants, electroplating activities, leather tanning, timber treatment, pulp production, mineral ore, and petroleum refining (Bolan et al. 2003a; Choppala et al. 2012). Chromium exists as hexavalent [Cr(VI)] and trivalent [Cr(III)] forms. The Cr(VI) is the characteristic of chromates, dichromates, and chromic trioxide, and the Cr(III) forms Cr oxides and hydroxides. While Cr(VI) is toxic and highly soluble in water, Cr(III) is less toxic, insoluble in water, and hence less mobile in soils (Barnhart 1997; Kosolapov et al. 2004). In soils, Cr exists mainly as Cr(III) unless oxidizing agents such as manganous oxide [Mn(IV)] are present (Gong and Donahoe 1997).

2.3 Mercury

The burning of fossil fuels and gold recovery in mining are the major sources of Hg (Pacyna et al. 2001; de Lacerda 2003). The chloralkali, electrical equipment, paint, and wood pulping industries are the largest consumers of Hg, accounting for 55% of the total consumption (Baralkiewicz et al. 2007). Mercury forms salts in two ionic states, mercurous [Hg(I)] and mercuric [Hg(II)], with the latter much more common in the environment than the former (Schroeder and Munthe 1998). These salts if soluble in water are bioavailable and toxic. Mercury also forms organometallic compounds, many of which have industrial and agricultural uses. Elemental Hg gives rise to a vapor that is only slightly soluble in water, but is problematic because of easy transport in the atmosphere (Boening 2000). The most common form of Hg mineral is insoluble mercuric sulfide (naturally occurring Cinnabar) which is nontoxic.

The environmental Hg cycle has four strongly interconnected compartments: atmospheric, terrestrial, aquatic, and biotic. The atmospheric compartment is dominated by gaseous Hg(0), although Hg(II) dominates the fluxes in the aquatic and terrestrial compartments. The terrestrial compartment is dominated by Hg(II) sorbed to organic matter in soils. The aquatic compartment is dominated by Hg(II)-ligand pairs in water and Hg(II) in sediments, and the biotic compartment is dominated by methyl Hg. Mercury is quite reactive in the environment and cycles readily among these compartments (Wiener et al. 2003).

2.4 Selenium

Among the elements, Se ranks 17th in order of abundance and is widely dispersed in the earth's crust in low concentration. Elemental Se exists as zero valence state and is often associated with sulfur in compounds such as selenium sulfide (SeS) and polysulfides. The principal sources of Se for commercial applications are Cu-bearing ore and sulfur deposits. Selenium is used in xerography, as a semiconductor in photocells, and also used in the manufacture of batteries, glass, electronic equipment, antidandruff products, veterinary therapeutic agents, feed additives, and fertilizers.

Selenium can be found in four different oxidation states: selenate [Se(VI); SeO_4^{2-}], selenite [Se(IV); SeO_3^{2-}], elemental selenium [Se(0); Se^0], and selenide [Se(−II); Se^{2-}]. Selenate and Se(IV) are common ions in natural waters and soils. Selenides and Se(0) are the common Se species in acidic soils that are under reducing conditions and are rich in organic matter. At moderate redox potential (Eh) either $HSeO_3^-$ or SeO_3^{2-} is the predominant form. At high Eh in well-aerated alkaline soils, the highly soluble SeO_4^{2-} is the predominant form.

Reduced Se compounds include volatile methylated species such as dimethyl selenide [DMSe, $Se(CH_3)_2$], dimethyl diselenide [DMDSe, $Se_2(CH_3)_2$], and dimethyl selenone [$(CH_3)_2 SeO_2$], and sulfur containing amino acids including selenomethionine, selenocysteine, and selenocystine. Inorganic reduced Se forms include mineral selenides and hydrogen selenide (H_2Se). Mosher and Duce (1987) estimated that approximately 90% of all natural emissions of Se to atmosphere are biogenic, with marine biosphere accounting for about 70% and the continental biosphere about 20%.

3   Microbial Transformation Processes

The microbial processes involved in transforming trace elements in soils and sediments are grouped into three categories that include bioaccumulation, oxidation/reduction, and methylation/demethylation (Fig. 1). Microorganisms can accumulate organometal(loid)s, a phenomenon relevant to toxicant transfer to higher organisms (i.e., biomagnification). In addition to bioaccumulating organometal(loid) compounds, many microorganisms are also capable of degrading and detoxifying them through various processes such as demethylation and dealkylation. Several organometal(loid) transformations are potentially useful for environmental bioremediation (Gadd 2010; Geoffrey and Gadd 2007).

3.1   Bioaccumulation

The physicochemical mechanisms by which trace elements are removed are encompassed by the general term "biosorption." Biosorption includes adsorption, ion exchange, entrapment, and metabolic uptake, which are features of both living and dead biomass and their derived products (Ahalya et al. 2003). In living cells, biosorption is directly and indirectly influenced by metabolism. Metabolism-dependent mechanisms of trace element removal that occur in living microorganisms include precipitation as sulfides, complexation by siderophores and other metabolites, sequestration by metal(loid)-binding proteins and peptides, transport

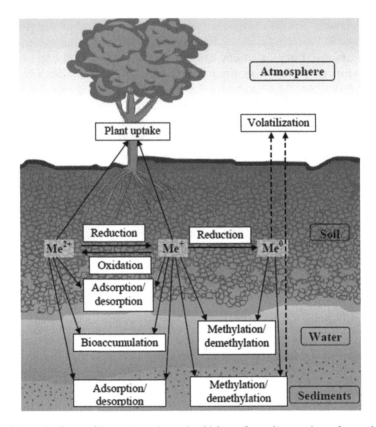

Fig. 1 Schematic diagram illustrating various microbial transformation reactions of trace elements in relation to remediation

and intracellular compartmentation (White et al. 1995). Microorganisms exhibit a strong ability to accumulate (bioaccumulation) trace elements from substrates containing very low concentrations (Robinson et al. 2006). Both bacteria and fungi bioaccumulate trace elements, and the bioaccumulation process is activated by two processes (Schiewer and Volesky 2000; Adriano et al. 2004): (1) sorption (i.e., biosorption) of trace elements by microbial biomass and its byproducts and (2) physiological uptake of trace elements by microorganisms through metabolically active and passive processes (Table 2).

Gram-positive bacteria possess cell walls that have powerful chelating properties. Trace element-binding to cell walls may be a biphasic process (Volesky and Holan 1995; Dostalek et al. 2004). The initial interaction between metal(loid) ions and reactive groups is followed by inorganic deposition of additional metal(loid)s. Trace elements can accumulate in greater than stoichiometric amounts, which cannot be solely accounted for by the ion-exchange processes. Metal(loid)-loaded bacterial cells have been shown to act as nuclei for the precipitation of crystalline metal(loid) deposits, when they are incorporated into contaminated sediments (Lawson and

Table 2 Selected references on the biosorption of arsenic, chromium, mercury, and selenium in soil and aquatic environments

Trace element	Medium	Biomass	Observation	Reference
Arsenic	Water	*Penicillium chrysogenum*	Amine-based surfactants or with a cationic polyelectrolyte increased biosorption of As(V)	Loukidou et al. (2003)
	Water	*Lessonia nigrescens*	As(V) achieved better sorption at low pH	Hansen et al. (2006)
	Water	Acid washed crab shells	As(V) bound to amide groups present in crab shells	Niu et al. (2007)
	Water	FeCl$_3$ pretreated tea fungal mat	Fe in fungal mat formed As(V)–Fe oxide bonds	Murugesan et al. (2006)
	Water	*Staphylococus xylosus*	As(V) and As(III) were adsorbed on the Fe(II) pretreated biomass surface through interaction with FeOH and FeOH$_2^+$ groups	Aryal et al. (2010)
	Soil	*Rhodococcus* sp. WB.12	Maximum sorption capacity of As(III) was 77.3 mg g^{-1} at 30°C, pH 7.0 and indicated involvement of several functional groups	Prasad et al. (2011)
Chromium	Tannery effluent	*Bacillus megaterium* and *B. coagulans*	Dead biomass sorbed more Cr(VI) than living cells	Srinath et al. (2002)
	Water	*Pinus sylvestris* cone biomass	pH of the aqueous phase strongly affected the sorption capacity of Cr(VI)	Ucun et al. (2002)
	Water	*Ocimum basilicum* seeds	Porous swollen outer layer of seeds increased sorption of Cr(VI)	Melo and D'Souza (2004)
	Water	*Sargassum* sp.	Maximum adsorption of Cr(III) was observed at 30°C, pH 3.5 and was 1.31 mmol g^{-1}	Cossich et al. (2004)
	Water	*Ecklonia* brown alga	Cr(III) adsorbs on –COOH functional groups	Yun et al. (2001)
	Water	*Ceramium virgatum*—red algae	Biosorption of Cr(III) and Cr(VI) was taken place by chemisorption	Tuzen and Sari (2010)

(continued)

Table 2 (continued)

Trace element	Medium	Biomass	Observation	Reference
Mercury	Water	*Ulva lactuca*—algae	Sorption depended on the pH of solution and concentration of Hg	Zeroual et al. (2003)
	Water	*Ricinus communis* leaves powder	Biosorption efficiencies increased with increasing contact time and initial metal(loid) concentration	Al Rmalli et al. (2008)
	Water	*Penicillium oxalicum var. Armeniaca, Tolypocladium* sp.	The greater efficiency may be due to deacetylation treatment used for cleaning the waste biomass	Svecova et al. (2006)
	Water	*Lessonia nigrescens* and *L. trabeculata*	Presence of Cl⁻ and competing ions such as Cd(II), Zn(II), and Ni(II) decreased Hg sorption	Reategui et al. (2010)
	Water	Magnetically modified *Saccharomyces cerevisiae* sub sp. *uvarum*	Biosorption of Hg^{2+} increased with an increase in pH and reached plateau at pH 5.5	Yavuz et al. (2006)
	Water	Estuarine *Bacillus* sp.	Changes in pH of solution had great effect on sorption of Hg on biomass	Green-Ruiz (2006)
Selenium	Water	Thiolated egg shell membranes	Thiol groups formed by modification increased biosorption of Se(IV) and Se(VI) species	Yang et al. (2010)
	Water	Wheat bran	Se(IV) and Se(VI) species were sorbed on biomass by ion-exchange process	Hasan et al. (2010)
	Water	Green algae—*Cladophora hutchinsiae*	Sorption of Se(IV) by biomass was through chemical ion exchange mechanism	Tuzen and Sari (2010)

Macy 1995; White et al. 1995). The metabolically independent passive sorption may account for the most significant portion of total uptake of a number of trace elements. For example, Surowitz et al. (1984) found up to 90% of total Cd uptake by *Bacillus subtilis* was located in the cell wall, 3–4% on the cell membrane, and the reminder in the soluble fraction of the cell. Similarly, 70–80% of Cu was accumulated as a layer on the cell wall of *Trichoderma viride* (Anand et al. 2006).

Biosorption of metal(loid) ions onto bacterial and fungal biomass is rapid and temperature dependent (Ledin et al. 1999; Dursun 2006). A wide range of binding groups, including carboxyl, amine, hydroxyl, phosphate (P), sulfhydryl have been shown to contribute to biosorption of trace elements. Both living and dead biomass act as biosorptive agents, and the magnitude of the phenomenon is directly related to these functional groups. Bacteria capable of producing large quantities of extracellular polymers are of a polysaccharide nature, have anionic properties, and are involved in removing soluble metal(loid) ions from solution by an ion-exchange process (Kodukula et al. 1994; Iyer et al. 2005). Macrofungi, such as *Agaricus* can bioaccumulate Cd and Hg from soils/compost containing low concentrations of these elements (Tüzen et al. 1998). Many fungal products, such as glucans, mannans, melanins, chitins, and chitosans have been shown to act as efficient biosorption agents (Gadd 1990). For example, biosorption of Cr(VI) by *Termitomyces clypeatus* occurred by initial rapid surface binding, followed by relatively slow intracellular accumulation through an active transport process. Chromate could bind to the surface of the cells by electrostatic interaction and be transported into the cytoplasm through different layers, including chitin–chitosan, glucan, and mannan of the cell wall and cytoplasmic membrane. Transported Cr(VI) was reduced to Cr(III) by enzymes present in the cytoplasm (Das and Guha 2009).

Several workers have examined the passive sorption of trace elements by both dead and living microorganisms, such as bacteria and fungi (Zouboulis et al. 2004; Mohanty et al. 2006; Das et al. 2008). For example, Ledin et al. (1999) used a multicompartment system, PIGS (Partitioning in Geobiochemical Systems), with five compartments to study metal(loid) distribution between soil constituents. Soil microorganisms (*Pseudomonas putida*, *Trichoderma harzianum*) were compared with common soil minerals (kaolin and aluminum oxide) and solid organic matter (peat), with respect to their ability to accumulate Zn, Cd, and Hg. Results indicated that for the different solid phases, metal(loid) distribution was related to variations in pH and ionic strength of the solutions. The presence of fulvic acid generally decreased metal(loid) accumulation by peat and microorganisms at near neutral pH values. The substantial accumulation of metal(loid)s by organic compounds (peat), as well as by microorganisms, especially under acidic conditions is of particular interest, since this process may counteract the metal(loid)-mobilizing effects of soil acidification.

Several trace elements are essential for many metabolic functions of microorganisms, and their uptake involves both active and passive processes (Zouboulis et al. 2004). In the case of metabolically active process, microorganisms exhibit specific mechanisms for the uptake of trace elements, which involve carrier systems associated with active ionic influxes across the cell membrane. Generally the metabolically active process is slower than passive absorption, requiring the presence of suitable energy source and ambient conditions. Within the cell, microorganisms may convert metal(loid) ions into innocuous forms by precipitation or binding. Microbially induced crystallization of metal(loid) is a well-known phenomenon (Gadd 2008). For example, sulfate-reducing bacteria such as *Desulfovibrio* release hydrogen sulfide (H_2S), thereby resulting in the precipitation of insoluble metal(loid) sulfides (Sass et al. 2009). Some bacteria produce siderophores in the form of phenols, catechols, or hydroxamates as part of their overall Fe uptake strategy (Renshaw et al. 2002; Hider and Kong 2010). Bacteria also produce metal(loid)-binding proteins, such as metallothionein that could bind metal(loid)s, thereby acting as detoxicants. For example, both *P. putida* and *Escherichia coli* have been shown to produce low-molecular weight proteins that bind Cd (Mejáre and Bülow 2001).

3.2 Redox Reactions

Trace elements, including As, Cr, Hg, and Se, are most commonly subjected to microbial oxidation/reduction reactions (Table 3). Redox reactions influence the speciation and mobility of trace elements. For example, metals generally are less soluble in their higher oxidation state, whereas the solubility and mobility of metalloids depend on both the oxidation state and the ionic form (Ross 1994). The oxidation/reduction reactions for various metal(loid)s and the optimum redox values for these reactions are given in Table 4. The redox reactions are grouped into two categories, assimilatory and dissimilatory (Brock and Madigan 1991). In assimilatory reactions, the metal(loid) substrate will serve a role in the physiology and metabolic functioning of the organism by acting as terminal electron acceptor, similar to oxygen (O_2) for aerobes and nitrate (NO_3^-) for denitrifying bacteria. In contrast, for dissimilatory reactions, the metal(loid) substrate has no known role in the physiology of the species responsible for the reaction and indirectly initiates redox reactions.

The biochemistry of microbial redox reactions of metal(loid)s has not been completely characterized (Miyata et al. 2007). In some systems (e.g., Se), metal(loid) transformation is coupled with the cytochrome system (Rosen and Silver 1987). Also specific metal(loid)-active enzymes may play a role in metal(loid) redox reactions. For example, *Thauera selenatis* reduces Se(IV) to Se(II) using selenate reductase (Debieux et al. 2011), and the subsequent reduction of Se(II) to Se(0) appears to be catalyzed by periplasmic nitrite reductase. However, Se reduction in a *Pseudomonas* sp. is part of the anaerobic respiration process.

Arsenic in soils and sediments can be oxidized to As(V) by bacteria (Battaglia-Brunet et al. 2002; Bachate et al. 2012; Table 3). Since As(V) is strongly retained by inorganic soil components, microbial oxidation results in the immobilization of As. Under well-drained conditions As would present as $H_2AsO_4^-$ in acidic soil and as $HAsO_4^{2-}$ in alkaline soils. Under reduced conditions, As(III) dominates in soils, but elemental arsenic [As(0)] and arsine (H_2As) can also be present. Arsenite is much more toxic and mobile than As(V). The distribution and mobilization of As species

Table 3 Selected references on the redox reactions of arsenic, chromium, mercury, and selenium in soil and aquatic environments

Trace element	Medium	Observation	Reference
Arsenic(V)	Soil	Dissimilatory As(V)-reducing bacterium, *Bacillus selenatarsenatis* increased the removal of As from contaminated soils	Yamamura et al. (2008)
	Soil	Both biotic and abiotic (S^{2-}, Fe^{2+}, and $H_2(g)$) factors were responsible for reduction of As(V)	Jones et al. (2000)
	Soil	Reduction mediated by H_2S was dominant in reducing conditions	Rochette et al. (2000)
	Soil	Attachment of *Shewanella putrefaciens* cells to oxide mineral surfaces promoted As(V) desorption, thereby facilitated its reduction	Huang et al. (2011)
Arsenic(III)	Soil	Fe(III) and Mn(III) oxides oxidized As(III) to As(V) through electron transfer reaction	Mahimairaja et al. (2005)
	Water	The oxidation rate of As(III) to As(V) and its subsequent adsorption was high in the presence of synthetic birnessite (MnO_2)	Manning et al. (2002)
	Water	In the presence of dissolved Fe(III) and near ultraviolet light, As(III) oxidized to As(V) in water	Emett and Khoe (2001)
	Water	As(III) oxidation rate increased in presence of *Euglena mutabilis*, a detoxification pathway	Casiot et al. (2004)
	Water	As(III) was oxidized to As(V) within hours in the presence of magnetite but reaction was quenched by the addition of ascorbic acid	Chiu and Hering (2000)
	Water	Bacteria attached to submerged macrophytes mediated the rapid As(III) oxidation reaction	Wilkie and Hering (1998)
	Water	Ferrihydrite oxidized As(III) to As(V) in the presence of light (photooxidation)	Bhandari et al. (2011)
Chromium(VI)	Soil	Application of Fe(II) under flow conditions increased reduction of Cr(VI)	Franco et al. (2009)
	Soil	Root exudates of *Typha latifolia* and *Carex lurida* increased sulfide species, which facilitated Cr(VI) reduction in sediment pore water	Zazo et al. (2008)
	Soil	Organic amendments increased DOC, which reduced Cr(VI) to Cr(III) in soils	Bolan et al. (2003a)
	Soil	Addition of glucose promoted both biotic and abiotic Cr(VI) reduction in soils	Leita et al. (2011)
	Water	Microbial fuel cells reduced Cr(VI) with the help of mixed consortium in autotrophic conditions	Tandukar et al. (2009)
	Water	Chitosan-FeO nanoparticles reduced Cr(VI) and pH of the solution played an important role	Geng et al. (2009)

(continued)

Table 3 (continued)

Trace element	Medium	Observation	Reference
Chromium(III)	Soil	Cr(III) was oxidized by atmospheric oxygen at high temperature to Cr(VI) in tannery sludge contaminated sites	Apte et al. (2006)
	Soil	Hydrous Mn(IV) oxides reacted with Cr(III) hydroxides and influenced the rate of Cr(III) oxidation	Landrot et al. (2009)
	Soil	In the alkaline soil with moderate organic C at the ore processing site, most of the Cr(VI) remained dissolved after H_2O_2 had decayed, indicating the mobilization of Cr(VI). As oxidation of organic C promotes disintegration of soil structure, it increases the access of solution to Cr(VI) mineral phases	Rock et al. (2001)
	Soil	The Mn oxide salts, birnessite, and todorokite had same capacity to oxidize Cr(III) to Cr(VI) in soils	Kim et al. (2002)
Mercury(II)	Water	Natural oxidation of Cr(III) to Cr(VI) by Mn (hydr)oxides formed by microbial activity	Ndung'u et al. (2010)
	Water	Carboxylic groups in the humic acids reduced Hg(II) to Hg(0)	Allard and Arsenie (1991)
	Water	Photochemical and humic mediated reduction of Hg(II) were two important components of the Hg(0) flux in the marine environment	Costa and Liss (1999)
	Water	Sulfite reduced Hg(II) to insoluble $HgSO_3$ and finally converted it to Hg(0)	Munthe et al. (1991)
	Water	Humic substances, especially sulfur containing ligands in terrestrial environment reduced Hg(II) to Hg(0)	Rocha et al. (2000)
Selenium(VI)	Soil and sediments	In suboxic conditions, green rust [Fe(II,III)] reduced Se(VI) to Se(0)	Myneni et al. (1997)
	Water	Microorganisms reduced Se(VI) to Se(IV) and reduction increased in the presence of lactate	Maiers et al. (1988)
	Soil	Organic amendments and low oxygen level in the soil reduced Se(VI) to Se(IV)	Guo et al. (1999)
	Soil and water	*Moraxea bovis* and bacterial consortia reduced Se(VI) to Se(IV) and Se(0)	Biswas et al. (2011)
	Water	*Bacillus pumilus* CrK08 had the highest capacity to reduce Se(VI) to Se(IV)	Ikram and Faisal (2010)
	Water	Nearly half of the total sorbed Se(IV) was reduced to Se(0) by Fe(II) sorbed on calcite within 24 h	Chakraborty et al. (2010)
Selenium(VI) and (IV)	Water	Rice straw, a good source of carbon and energy, helped several bacteria in reducing Se(VI) and Se(IV) to Se(0)	Frankenberger et al. (2005)
	Water	Formic acid and methanol/ethanol promoted photoreduction of Se(VI) and Se(IV) to Se(0) in the presence of TiO_2	Tan et al. (2003)
Se(0)	Sediments	*Bacillus selenitireducens* and some other bacteria were capable to reduce Se(0) to Se(II) in anoxic sediments	Herbel et al. (2003)

Table 4 Oxidation–reduction reactions of selected elements and the optimum redox potentials

Trace element	Transformation	Reaction	E_0 (mV)
Arsenic(0)	$As(0) + 3H_2O \rightarrow H_3AsO_3 + 3H^+ + 3e^-$	Oxidation	+250
Arsenic(V)	$H_3AsO_4 + 2H^+ + 2e^- \rightarrow H_3AsO_3 + H_2O$	Reduction	+560
Cadmium(0)	$Cd \rightarrow Cd^{2+} + 2e^-$	Oxidation	+402
Cadmium(II)	$Cd^{2+} + 2e^- \rightarrow Cd$	Reduction	−400
Chromium(III)	$2Cr^{3+} + 3H_2O + 2MnO_4^- \rightarrow 2Cr_2O_7^{2-} + 6H^+ + 2MnO_2$	Oxidation	+350
Chromium(VI)	$Cr_2O_7^{2-} + 14H^+ + 6e^- \rightarrow 2Cr^{3+} + 7H_2O$	Reduction	+1,360
Iron(II)	$Fe^{2+} + 2e^- \rightarrow Fe(s)$	Oxidation	−440
Iron(III)	$Fe^{3+} + e^- \rightarrow Fe^{2+}$	Reduction	+770
Manganese(II)	$Mn^{2+} + 2e^- \rightarrow Mn$	Reduction	−1,180
Manganese(IV)	$Mn^{4+} + 2e^- \rightarrow Mn^{2+}$	Reduction	+1,210
Manganese(VII)	$MnO_4^- + 8H^+ + 5e^- \rightarrow Mn^{2+} + 4H_2O$	Reduction	+1,510
Manganese(VII)	$MnO_4^- + e^- \rightarrow MnO_4^{2-}$	Reduction	+564
Mercury	$Hg_2^{2+} + 2e^- \rightarrow 2Hg$	Reduction	+790
Nitrogen(V)	$2NO_3^- + 4H^+ + 2e^- \rightarrow 2NO_2 + 2H_2O$	Reduction	+803
Selenium(0)	$Se(0)/H_2Se$	Reduction	−730
Selenium(VI)	SeO_4^{2-}/SeO_3^{2-}	Reduction	+440
Selenium(VI)	$SeO_3^{2-}/Se(0)$	Reduction	+180
Sulfur(II)	$S^{2-} + 2H^+ \rightarrow H_2S$	Reduction	−220
Sulfur(VI)	$SO_4^{2-} + H_2O + 2e^- \rightarrow SO_3^{2-} + 2OH^-$	Reduction	−520

in the soil and sediments is controlled by both microbially mediated transformation of the As species and by adsorption (Adriano et al. 2004; Mahimairaja et al. 2005). The reduction and methylation reactions of As in sediments are generally mediated by bacterial degradation of organic matter coupled with reduction and use of sulfate as the terminal electron acceptor (Adriano et al. 2004). Ferrous iron [Fe(II)] can also serve as an electron acceptor in bacterial oxidization of organic matter, resulting in the decomposition of ferric [Fe(III)] oxides and hydroxides.

Although Cr(III) is strongly retained on soil particles, Cr(VI) is very weakly adsorbed in soils that are net negatively charged and is readily available for plant uptake and leaching to groundwater (James and Bartlett 1983; Leita et al. 2011). Oxidation of Cr(III) to Cr(VI) is primarily mediated abiotically through oxidizing agents such as Mn(IV), and to a lesser extent by Fe(III), whereas reduction of Cr(VI) to Cr(III) is mediated through both abiotic and biotic processes (Choppala et al. 2012). Oxidation of Cr(III) to Cr(VI) can enhance the mobilization and bioavailability of Cr. Chromate can be reduced to Cr(III) in environments where a ready source of electrons (Fe(II)) is available. Suitable conditions for microbial Cr(VI) reduction occur where organic matter is present to act as an electron donor, and Cr(VI) reduction is enhanced under acid rather than alkaline conditions (Hsu et al. 2009; Chen et al. 2010).

In living systems, Se tends to be reduced rather than oxidized, and reduction occurs under both aerobic and anaerobic conditions. Dissimilatory Se(IV) reduction to Se(0) is the major biological transformation for remediation of Se oxyanions in anoxic sediments (Lens et al. 2006). Selenite is readily reduced to the elemental state by chemical reductants such as sulfide or hydroxylamine, or biochemically by systems such as glutathione reductase. Hence, precipitation of Se in its elemental form, which has been associated with bacterial dissimilatory Se(VI) reduction, has great environmental significance (Oremland et al. 1989, 2004). Since both Se(IV) and NO_3^- can be used as terminal electron acceptors by many microorganisms, the presence of NO_3^- in the system inhibits the reduction of Se(IV) (Viamajala et al. 2002).

Mercury undergoes a biological reduction process in soils and sediments. Microorganisms play a major role in reducing reactive Hg(II) to nonreactive Hg(0), which may be subjected to volatilization losses. Bacteria are more important than eukaryotic phytoplankton in the reduction of Hg(II). Mercury-resistant bacteria can transform ionic mercury [Hg(II)] to metallic mercury [Hg(0)] by enzymatic reduction (Von Canstein et al. 2002). It is known that Hg(II) is reduced to Hg(0) by mercuric reductase, a mercury resistance operon, and genetic system encoding transporters and regulators. The dissimilatory metal(loid) reducing bacterium *Shewanella oneidensis* has been shown to reduce Hg(II) to Hg(0), which requires the presence of electron donors (Wiatrowski et al. 2006).

3.3   Methylation/Demethylation

Methylation has been proposed as a mechanism for detoxification—a biological mechanism for the removal of toxic trace elements by converting them to methyl derivatives that are subsequently removed by volatilization or extraction with solvents (Frankenberger and Losi 1995). Methylated derivatives of As, Hg, and Se can arise as a result of chemical and biological mechanisms, and this frequently results in altered volatility, solubility, toxicity, and mobility. The major microbial methylating agents are methylcobalamin (CH_3CoB_{12}), involved in the methylation of Hg, and *S*-adenosylmethionine (SAM), involved in the methylation of As and Se. Biological methylation (biomethylation) may result in metal(loid) detoxification, since methylated derivatives may be excreted readily from cells and are often volatile and may be less toxic, e.g., organoarsenicals. However, for Hg, methylation may not play a major role in detoxification, because of the existence of more efficient resistance mechanisms, e.g., reduction of Hg(II) to Hg(0) (Gadd 1993).

Volatilization occurs through microbial conversion of metal(loid)s to their respective metallic, hydride, or methylated forms. These forms have low boiling points and/or high vapor pressure, and are therefore susceptible for volatilization (Table 5). Methylation has shown to be the major process by which As, Hg, and Se are volatilized from soils and sediments, and which also releases poisonous methyl gas (Adriano et al. 2004).

Although methylation of metal(loid)s occurs through both chemical (abiotic) and biological processes, biomethylation is considered to be the dominant process in soils and aquatic environments (Table 6). Thayer and Brinckman (1982) grouped methylation into two categories: trans-methylation and fission-methylation. Trans-methylation refers to the transfer of an intact methyl group from one compound (methyl donor) to

Table 5 Boiling points and vapor pressure of elemental, hydride, and methyl species of arsenic, mercury, and selenium

Trace element	Species	Boiling point (°C)	Vapor pressure (kPa)
Arsenic	Elemental—As(0)	603	0 (Approx)
	Hydride—AsH	−55	1,461.2 at 21.1°C
	Methylated:		
	Monomethyl arsine—As(CH$_3$)	1.19	231.9 at 25°C
	Dimethyl arsine—As(CH$_3$)$_2$	36	68.26 at 25°C
	Trimethyl arsine—As(CH$_3$)$_3$	53.8	42.92 at 25°C
Chromium	Elemental—Cr(0)	2,672	0.99 at 1,857°C
Mercury	Elemental—Hg(0)	356.58	2×10^{-7} at −38.72°C
	Hydride—HgH$_2$	−34	–
	Methylated:		
	Monomethyl mercury Hg(CH$_3$)	–	–
	Dimethyl mercury Hg(CH$_3$)$_2$	93.5	–
Selenium	Elemental—Se(0)	688	6.95×10^{-4} at 221°C
	Hydride—SeH$_4$	−42	–
	Methylated:		–
	Dimethyl selenide—Se(CH$_3$)$_2$	49.8	41.1 at 25°C
	Dimethyl selenite—Se$_2$(CH$_3$)$_2$	−39.042	826.85 at 25°C
	Dimethyl selenone—Se(CH$_3$)$_2$O$_2$	153	–

another compound (methyl acceptor). Fission-methylation refers to the fission of a compound (methyl source), not necessarily containing a methyl group, so as to eliminate a molecule such as formic acid. The fission molecule is subsequently captured by another compound which is reduced to a methyl group.

At present there is substantial evidence for the biomethylation of As, Hg, and Se in soils and aquatic systems (Table 6). Microorganisms in soils and sediments act as biologically active methylators (Frankenberger and Arshad 2001; Loseto et al. 2004). Organic matter provides the methyl-donor source for both biomethylation and abiotic methylation in soils and sediments. Methylation of Hg is controlled by low molecular weight fractions of fulvic acid in soils. Using a synthetic humic acid fraction, Ravichandran (2004) concluded that the methylating factor was not associated with the higher molecular weight humic acid fraction. Similarly, Lambertsson and Nilsson (2006) suggested that organic matter content and the supply of alternative electron acceptors influenced methylation of Hg in the sediments. Organic matter may affect methylation of Hg through several mechanisms including maintenance of low Eh and providing electron and complexing agents.

Biomethylation is effective in forming volatile compounds of As such as alkylarsines, which could easily be lost to the atmosphere (Lehr 2003; Yin et al. 2011). Methylation, demethylation, and reduction reactions are important in controlling the mobilization and subsequent distribution of arsenicals. These transformations are promoted by microbes although it is still not clear if in situ biomethylation is a common phenomenon.

Arsenic undergoes a series of biological transformation in aquatic systems, yielding a large number of compounds, especially organoarsenicals (Maher and Butler 1988).

Table 6 Selected references on the methylation/demethylation of arsenic, mercury, and selenium in soils

Trace element	Organisms	Observations	Reference
Arsenic	Cyanobacterium sp. (Synechocystis sp. and Nostoc sp.)	Methylated As(III) to less toxic trimethylarsine as an end product in water and soil	Yin et al. (2011)
	Cyanidioschyzon sp. (Eukaryotic alga)	Methylated As(III) to form trimethylarsine oxide and dimethylarsenate	Qin et al. (2009)
	Streptomyces sp.	Demethylated methylarsine acid to As(III) species	Yoshinaga et al. (2011)
	Ecklonia radiata	Arsenosugars were degraded to As(V) with dimethylarsinoylethanol and dimethylarsinate as intermediate species and observed that non-extractable, recalcitrant As increased in decaying algae	Navratilova et al. (2011)
	Methanobacterium archaea	Methanobacterium sp. cultures were most effective in producing volatile As derivatives	Michalke et al. (2000)
Mercury	Polygonum densiflorum	The polygonum sp. found in periphyton associated to macrophyte roots may be the dominating Hg methylating family among sulfate reducing bacteria	Achá et al. (2011)
	Desulfomicrobium escambiense	Desulfomicrobium sp. had the highest methylating capability of Hg (45%), and methylation was strain specific but not species- or genus–dependent	Bridou et al. (2011)
	Desulfovibrio desulfuricans	D. desulfuricans ND132 strain has the ability to produce methyl Hg and its further degradation. However, the presence of sulfide decreased the methylation process	Gilmour et al. (2011)
	Geobacter sp.	Geobacter sp. strain CLFeRB, an iron reducing bacterium methylated Hg at rates comparable to sulfate reducing bacteria	Fleming et al. (2006)
Selenium	Escherichia coli	Escherichia cells encoding the bacterial thiopurine methyltransferase have methylated Se(IV) and methylselenocysteine into DMSe and DMDSe	Ranjard et al. (2002)
	Enterobacter cloacea	Reduced toxic oxyanions of Se to insoluble Se(0) and Se biomethylation was protein/peptide-limited process	Frankenberger and Arshad (2001)
	Acremonium falciforme	Methylated inorganic Se more rapidly than organic forms	Thompson-Eagle et al. (1989)
	Alternaria alternate (black microfungi)	Volatilized Se to DMSe and DMDSe even after the immobilization by goethite	Peitzsch et al. (2010)

Benthic microbes are capable of methylating As under both aerobic and anaerobic conditions to produce methylarsines and methyl-arsenic compounds with a generic formula $(CH_3)_n As(O)(OH)_{3-n}$ where n may be 1, 2, or 3. Monomethyl arsenate (MMA) and dimethyl arsinate (DMA) are common organoarsenicals in river water. Methylated As species could result from direct excretion by algae or microbes or from degradation of the excreted arsenicals or more complex cellular organoarsenicals (Li et al. 2009). Methylation may play a significant role in the mobilization of As by releasing it from the sediments to aqueous environment (Anderson and Bruland 1991; Wang and Mulligan 2006). In the sediments-aquatic system, methylation occurs only in the sediments because the thermodynamics of water or aquatic environments are not favorable for methylation (Duester et al. 2008). The rate of methylation/demethylation reactions and the consequent mobilization of arsenicals are affected by adsorption by sediments and soils. Adsorption of As species in alkaline condition follow: As(V) > As(III) > As(II) > DMA (Bowell 1994).

Primary producers, such as algae take up As(V) from solution and reduce it to As(III), prior to methylating the latter to produce MMA and DMA; the methylated derivatives are then excreted (Knauer et al. 1999). This may represent a detoxification process for the organism. Arsenic is taken up by algae due to its chemical similarity to P (Elbaz-Poulichet et al. 2000). Although the detoxification of As can be achieved through methylation, the element may be significantly toxic to phytoplankton and periphyton communities in marine environments.

Mercury is methylated through biotic and abiotic pathways, although microbial methylation mediated mainly through dissimilatory sulfate- and iron-reducing bacteria is generally regarded as the dominant environmental pathway (Musante 2008; Graham et al. 2012). Methylation of Hg occurs under both aerobic and anaerobic conditions (Rodríguez Martín-Doimeadios et al. 2004). Regnell and Tunlid (1991) showed that the proportion of methylated Hg was significantly higher in the anaerobic condition than in aerobic systems consisting of undisturbed lake sediment and water. The sequence of volatility of Hg species is as follows: $Hg(0) \approx (CH_3)_2 Hg >>> CH_3 HgCl > Hg(OH)_2 > CH_3 HgOH > HgCl_2$.

Under anaerobic conditions Hg(II) ions can be biologically methylated to form either monomethyl or dimethyl Hg. Monomethyl or dimethyl Hg are highly toxic (neurotoxic) and more biologically mobile than the other forms (Adriano et al. 2004). The main methylation mechanism for Hg involves nonenzymatic transfer of methyl groups of methylcobalamin (a vitamin B_{12} derivative, produced by many microorganisms) to Hg(II) ions (Ullrich et al. 2001; (1) and (2)). Methylation occurs both enzymatically and nonenzymatically; ionic species Hg(II) are required for the proceeding of biological methylation of Hg to proceed.

$$Hg(II) + 2R - CH_3 \rightarrow CH_3 HgCH_3 \rightarrow CH_3 Hg^+. \tag{1}$$

$$Hg(II) + R - CH_3 \rightarrow CH_3 Hg^+ + R - CH_3 \rightarrow CH_3 HgCH_3. \tag{2}$$

Methylmercury degradation occurs either through microbial or photochemical processes. Marvin-Dipasquale et al. (2000) observed that methyl Hg degrades mainly through reduction-degradation and oxidative-demethylation pathways in

Hg-contaminated sediments. Singlet O_2, a highly reactive form of dissolved O_2, produced upon sunlight irradiation (ultraviolet spectrum (<400 nm)) of dissolved organic matter, is primarily responsible for methyl Hg photochemical degradation (Zhang and Hsu-Kim 2010). However, the photodegradation is mainly dependent on the ligands that bind methyl Hg cations in water. Photodegradation of methyl Hg is higher in fresh water lakes and glaciers than in sea waters (Hammerschmidt and Fitzgerald 2006; Whalin et al. 2007).

Unlike As, Se(VI) is more mobile than Se(IV), because the former is strongly adsorbed onto soil minerals and organic matter under near neutral pH conditions (Li et al. 2008). Soluble Se species such as Se(VI) are unlikely to be found under reducing conditions because less soluble forms such as Se(0) are thermodynamically favored. When Se(IV) and Se(VI) are introduced into moderately reducing conditions they are readily transformed through microbial processes to Se(0) and/or organic Se compounds. Five different volatile forms of reduced Se have been detected: H_2Se, methane selenol (CH_3SeH), DMSe, dimethyl selenenyl sulfide (CH_3SeSCH_3), and DMDSe (Wu 2004). The relatively high vapor pressure of these compounds enhances the transformation of Se from soils and sediments to aqueous and vapor phases. However, the rapid oxidation of the first two compounds and lower vapor pressure of the last two leave DMSe as the most significant contributor to atmospheric Se input (Frankenberger and Losi 1995).

Selenium biomethylation is of interest because it represents a potential mechanism for removing Se from contaminated environments, and it is believed that methylated compounds, such as DMSe is less toxic than dissolved Se oxyanions. Fungi are more active in the methylation of Se in soils although some Se-methylating bacterial isolates have also been identified (Adriano et al. 2004). Hydrogen oxidizing methanogens such as *Methanobacterium omelianskii* are involved in the reductive methylation, while methylotrophic bacteria carry out demethylation. Dimethyl selenide can be demethylated in anoxic sediments as well as anaerobically by an obligate methylotroph similar to *Methanococcides methylutens* in pure culture.

An anaerobic demethylation reaction may result in the formation of toxic and reactive H_2Se from less toxic DMSe. Although H_2Se undergoes rapid chemical oxidation under toxic conditions, it can exist for long periods in an aerobic environment (Tarze et al. 2007). Because demethylation produces CO_2 in addition to CH_4, it is preceded by oxidative pathways used in substrate metabolism, rather than by lyases. Aerobic demethylation of DMSe is likely to yield Se(VI), thereby retaining Se in the system.

4 Factors Affecting Microbial Transformation Processes

Microbial transformation of As, Cr, Hg, and Se in soils, sediments, and aquatic systems is affected by biological functioning of the system, as measured by microbial activity, the bioavailability of the metal(loid) ions, as measured by speciation, and the physicochemical characteristics of the media such as pH, moisture content, and temperature. The microbial transformation processes can also be manipulated through the addition of inorganic and organic amendments (Park et al. 2011a).

4.1 Microbial Activity

Although the oxidation/reduction and methylation/demethylation reactions occur through both chemical and biological processes, the biological process is considered to predominate in soils, sediments, and aquatic systems, especially in the absence of adequate levels of inorganic sources of electron donor/acceptor such as Fe^{2+} and Mn^{2+} ions. For example, Losi et al. (1994), Bolan et al. (2003a), and Choppala et al. (2012) have shown that the addition of organic manure caused a greater increase in the biological reduction than the chemical reduction of Cr(VI), which suggests that the supply of microorganisms is more important than the supply of organic carbon (C) in the manure-induced Cr(VI) reduction. Addition of organic manure has often been shown to increase the microbial activity of soils through increased supply of both C and nutrient sources (Wardle 1992; Marinari et al. 2000).

Rogers (1976) and Song and Heyst (2005) observed that, in general, although the initial rate of volatilization through methylation of Hg was greater from clay than sandy soils, the clay soil resulted in less total loss of Hg. The greater initial rate of methylation in the clay soil may be attributed to the larger population of microorganisms. However, the low volatilization of Hg from the clay soil may be related to the greater adsorption of Hg resulting in the inaccessibility of Hg for microorganisms. The relative amounts of mono and dimethyl Hg formed depends on microbial species, organic pollutant loading, Hg concentration, and temperature and pH of the system.

Availability of nutrients and water, pH, and soil texture are significantly different with soil depth, thereby influencing the structure of microbial communities in soil. Anaerobic environments within soil aggregates also affect anaerobic- and aerobic-based metabolism by a microbial community (Hansel et al. 2008). Therefore, redox or methylation reactions may vary with soil profile, because these reactions are related to microbial activity.

4.2 Solute Concentration and Speciation

The rate of microbial transformation of trace elements depends on the bioavailability of the metal(loid) concerned, as measured by its concentration and speciation. Watras and Bloom (1992) reported that the bioaccumulation of methyl Hg results from two processes: the higher affinity of inorganic Hg in lower tropic level organisms and the high affinity of methyl Hg in fishes. Mason et al. (1995) compared the bioaccumulation of inorganic Hg and methyl Hg and showed that passive uptake of the Hg complexes ($HgCl_2$ and CH_3HgCl) results in high concentration of both the

Fig. 2 Relationship between reactive mercury and organic mercury (Bisinoti et al. 2007)

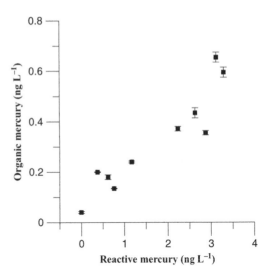

inorganic and methylated Hg in phytoplankton. However, differences in portioning within phytoplankton cells between inorganic Hg (which is principally membrane-bound) and methyl Hg (which is accumulated in the cytoplasm) lead to a greater assimilation of methyl Hg during zooplankton grazing.

Several environmental factors influence methylation rates of Hg, most probably by altering its bioavailability. Reactive and free Hg ions are required to initiate microbial transformation reactions involving microorganisms (Fig. 2). For example, complexing of Hg with chlorine and bromine decreases volatilization due to the nonavailability of free Hg radicals to produce volatile methyl Hg compounds by microorganisms (Rogers and MacFarlane 1978). Similarly complexation of Hg with dissolved organic carbon (DOC) decreases methylation (Miskimmin et al. 1992; Ravichandran 2004). Release of HgS during sulfate reduction inhibits methylation, whereas precipitation of S with Fe increases Hg available for methylation (Choi and Bartha 1994).

At low Hg concentration most of the Hg ends up as dimethyl Hg, whereas at high Hg concentration some of the Hg may remain as monomethyl Hg (Adriano 2001). The volatilization loss of methyl Hg depends on the nature of the methyl compound; dimethyl Hg is more volatile than monomethyl Hg. The speciation of Hg in water is most strongly influenced by the aqueous chemical conditions—most notably Eh, pH, and organic and inorganic ligands. Inorganic Hg(II) and methyl Hg are strongly influenced by the chemical makeup of the water and almost entirely form ion pairs with ligands in the aquatic environment. The complexation of Hg(II) with sulfide has been shown to strongly affect the availability of Hg for methylation by microbes (Benoit et al. 1999; Drexel et al. 2002).

To be methylated by bacteria, Hg(II) must first cross the cell membrane of a methylating bacterium, presumably as a neutral dissolved species. Thus, speciation of inorganic Hg in aqueous and solid phases controls the fraction of the total Hg pool that is available for microbial methylation. For example, a low concentration of chloride and sulfide ligands increases the bioavailability of Hg by forming neutral species such as $HgCl_2$ or HgS; however, at higher ligand concentrations, these ion pairs become charged (e.g., $HgCl_3^-$), thereby resulting in the adsorption of Hg and decreasing the supply of Hg for methylation (Boszke et al. 2003). Similarly, when Hg(II) is bound to large molecules of DOC it will become unavailable for methylation (Ravichandran 2004).

The reduction of Se by bacteria has often been shown to be affected by the initial Se concentration. For example, Lortie et al. (1992) observed that Se reduction rate by *Pseudomonas stutzeri* increased with increasing Se(VI) and Se(IV) concentrations up to 19 mM. At Se concentration more than 19 mM Se(IV) reduction decreased, while Se(VI) reduction remained constant, which might be attributed to the higher toxicity of the Se(IV) than Se(VI). Negatively charged Se oxyanions form ternary Se–cation–organic matter complexes and result in the immobilization of Se. It is known that Se oxyanions are reduced to elemental Se or Se(−II) and are removed by precipitation from solution; however, Se(VI) is kinetically stable in groundwaters that are high in humic substances and is more resistant to reduction (Fernández-Martínez and Charlet 2009).

When the organic matter contents are high in shallow subsurface environments, reducing conditions with a relatively high amount of sulfide in the solid phase decrease dissolved As concentration in the pore water. Arsenic is sorbed with the formation of Fe sulfides. Under oxidizing conditions, surface waters are undersaturated with As(V) mineral, and as a result, secondary As(V) dissolves and As concentration increases in the water. A close relationship of the As(III)/As(V) ratio with DOC concentration indicates that microbial processes play a key role in the transformation of As species (Drahota et al. 2009). Various anions influence As fractionation and mobility in fine fraction soil. Arsenic mobility in the fine soil increased in the order of $PO_4^{3-} >> CO_3^{2-} > SO_4^{2-} = Cl^-$ anions. Arsenic mobilization by PO_4^{3-} may be caused by a ligand exchange mechanism (Goh and Lim 2005).

Choppala (2011) observed that the addition of Fe(III) oxide to Cr(VI)-contaminated soils resulted in a decrease in the rate of reduction of Cr(VI) as measured by half-life values (Fig. 3a). This phenomena may be due to the increased retention of Cr(VI) by Fe(III) oxide, thereby decreasing the bioavailability of Cr(VI) for microorganisms. Similarly, an increase in pH decreased Cr(VI) reduction in soil (Fig. 3b). Both protons and electrons are needed for the reduction of Cr(VI) to Cr(III) (Park et al. 2004). An increase in pH decreases the H^+ ions in the soil, thereby decreasing Cr(VI) reduction.

4.3 Soil pH

Soil pH affects microbial transformation processes through its effects on the microorganisms, supply of protons, and adsorption and speciation of metal(loid)s. For example, Kelly et al. (2003) and Roy et al. (2009) observed that the extent of

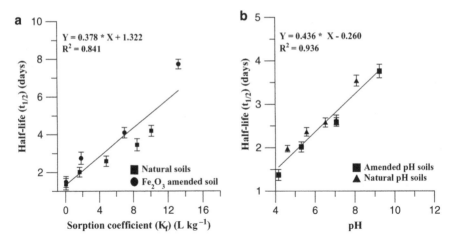

Fig. 3 Relationship between (**a**) chromate (Cr(VI)) adsorption as measured by K_f values and Cr(VI) reduction as measured by half-life, (**b**) pH and half-life for Cr(VI) reduction (Choppala 2011)

methylation decreased as soil pH increased. This has been attributed to the unavailability of Hg(II) at high pH, from its stronger adsorption and to the reduced supply of methylating organic matter at high pH. Neutral and alkaline environments favor dimethyl Hg; under acidic conditions, less volatile monomethyl Hg is formed, reducing the release of volatile Hg to the atmosphere.

pH can also influence bioaccumulation of trace elements by microorganisms. An increase in pH facilitates surface binding of trace elements (Yin et al. 2002) through pH-induced increases in surface charge. Furthermore, energy-dependent trace element uptake is frequently pH dependent (Öztürk et al. 2004) and maximum rates are observed between 5 and 7 (Sağlam et al. 1999; Bishnoi et al. 2007).

pH has been shown to influence the methylation rate by controlling organic C compounds with functional groups that would otherwise bind Hg (Adriano 2001; Gilmour and Henry 1991). Protons are required for reducing Cr(VI) to Cr(III) (3). An increase in Cr(VI) reduction after manure addition was noted by Losi et al. (1994) and Bolan et al. (2003a), and may also be attributed to the supply of protons (Fig. 3b). Organic manure is generally rich in N, part of which is in ammoniacal form. Oxidation of ammoniacal N to NO_3^- N (nitrification) and subsequent ammonia volatilization produce protons. Similarly, Camargo et al. (2003) found that Cr-resistant bacteria (*Bacillus* sp.) tolerated 2,500 mg L^{-1} Cr(VI), and the maximal Cr(VI) reduction occurred at pH 7.0–9.0. It has often been observed that Cr(VI) reduction, being a proton consumption (or hydroxyl release) reaction, increases as soil pH decreases (Eary and Rai 1991; Choppala et al. 2012; Fig. 3b).

$$2Cr_2O_7^{2-} + 3C^0 + 16H^+ \rightarrow 4Cr(III) + 3CO_2 + 8H_2O. \tag{3}$$

Mercury methylation decreased with decreasing pH of sediment, and methylation was not detected at a pH value< 5.0, which may be related to the unavailability of inorganic Hg (Ramial et al. 1985). Baker et al. (1983) reported that the methylation of Hg from inorganic mercuric chloride occurred in a narrow pH range of 5.5–6.5, perhaps because the microbial population had been adapted to a pH of 5.8 of the tested sediment. However, methyl arsenic acid and dimethyl arsenic acid were formed in the pH range of 3.5–7.5, because the As-methylating microorganisms were not sensitive to changes in pH (Baker et al. 1983). Kelly et al. (2003) investigated the influence of increasing protons on the uptake of Hg by an aquatic bacterium. A small decrease in pH (7.3–6.3) significantly increased Hg uptake by bacteria. Changes in Hg uptake by bacteria also affected Hg cycling, including elemental Hg production, Hg sedimentation, and methylation of Hg.

Fulladosa et al. (2004) studied the effect of pH on As(V) or As(III) speciation, and the resulting toxicity was investigated using the Microtox® bioassay (Azur Environmental Ltd, 1998), based on change in light emission by the luminescent bacteria *Vibrio fischeri*. Within a 5.0–8.0 pH range, EC_{50} values for As(V) were found to decrease as pH became basic, reflecting an increase in toxicity; for As(III), the EC_{50} values were almost unchanged within a 6.0–8.0 pH range and were lowered only at pH 9.0. They observed that the highest toxicity to *V. fischeri* occurred at basic pH values when $HAsO_4^{2-}$ and H_2AsO^{3-} species representing As(V) and As(III) were predominant.

Thompson-Eagle et al. (1989) showed that DMSe production by *Alternaria alternata* was affected by the reaction mixture pH. The optimum pH for methylation of Se was 6.5 (Frankenberger and Arshad 2001). Dissolved Se(VI) constituted 95% of the total soluble Se at pH 9 and decreased to 75% at pH 6.5. pH greatly influenced Se speciation, solubility, and volatilization, therefore indicating that pH is an important factor in the Se biogeochemistry (Masscheleyn et al. 1990).

4.4 Soil Moisture and Aeration

Soil moisture influences redox reactions by controlling the activity of microorganisms and also redox conditions of the microenvironment. An increase in moisture content and the amount of available C tend to increase the net loss of methyl Hg (Schlüter 2000; Oiffer and Siciliano 2009). Air-drying soil inhibits methylation of Se, while saturation with water causes anaerobiosis, thus decreasing the transfer of volatile Se from soil to air (Calderone et al. 1990). Alternate wetting and drying enhances the release of volatile Se compounds, which is attributed to release of nutrients through organic matter mineralization (Hechun et al. 1996).

Most studies showed that irrespective of the matrix type, greater quantities of volatile Se are released under aerobic than anaerobic conditions (Thompson-Eagle and Frankenberger 1990). For example, soil, sewage sludge, and seleniferous pond water samples released greater quantities of volatile Se when exposed to air than to N_2. Soil physical properties including aggregation and moisture content influence

Fig. 4 Relationship between soil water potential and selenium volatilization in a soil-pickle weed (*Salicornia bigelovii*) system (Shrestha et al. 2006)

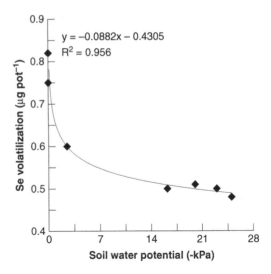

Eh of soil and aeration-dependent microbial activities which are important to nutrient cycling, soil fertility, and biogeochemistry of trace elements (Skopp et al. 1990; Fig. 4). When the soil moisture content is high, the diffusion of O_2 is limited and local anaerobic environment is created. In aerobic soils the predominant As species is As(V) in soil pore water, while As(III) comprised up to 80% of the total As in anaerobic soils. Chemical reducing conditions increases As(III) in anaerobic soils. Chemical conversion of As(V) to As(III) may reduce microbial activity and production of monomethyl arsenic acid (Haswell et al. 1985).

The methylation rate for Hg was found to be higher under anaerobic sediment conditions than under aerobic condition; anaerobic conditions inhibited the conversion of methyl Hg. More than 90% of the methyl Hg was formed by biochemical processes under anaerobic conditions (Berman and Bartha 1986; Akagi et al. 1995). The prevalence of methylcobalamin and a natural methylating agent that transfers methyl groups under anaerobic conditions may facilitate the methylation of inorganic Hg (Ullrich et al. 2001).

4.5 Soil Temperature

Temperature influences microbial transformation reactions of trace elements, mainly by controlling microbial activity and functions (Alexander 1999). Schwesig and Matzner (2001) and Heyes et al. (2006) observed that both the production and the volatilization loss of methyl Hg were directly proportional to temperature. Kocman and Horvat (2010) performed a laboratory-based experimental study of Hg emission from contaminated soils in the River Idrijca catchment (Slovenia) that regulated three major environmental parameters, comprising soil temperature (0 and 25°C, measured at the depth of 5 mm), soil moisture, and solar radiation. They noticed a strong positive correlation between the soil surface temperature and Hg emission flux. They suggested that this thermally controlled emission of Hg from soils depended on the equilibrium of Hg(0) between the soil matrix and the soil gas. As suggested by Schlüter (2000), because of increasing thermal motion, the vapor pressure of highly volatile Hg(0) is increased, and sorption by soil is decreased. Moreover, increased temperature also caused an increase in reaction rates and microbiological activity, resulting in more intensive formation of volatile Hg species (Schlüter 2000; Zhang et al. 2001; Bahlmann et al. 2006).

Temperature is one of the most important environmental factors that affect the rate of Se volatilization (Frankenberger and Karlson 1994a). For every 10°C increase in the temperature, the vapor pressure of volatile Se raises three- to four-fold (Karlson et al. 1994). Gao and Tanji (1995) have developed models to show that the methylation of Se in aquatic systems increases with both temperature and the supply of C as a source of methyl donor. Dungan and Frankenberger (2000) observed that the optimum temperature for Se volatilization was 35°C, with the rate of Se volatilization increasing as the temperature increased from 12 to 35°C. At 40°C, the rate of Se volatilization was slightly less than 35°C, but greater than at 30°C.

Camargo et al. (2003) found that Cr(VI) reduction by a Cr-resistant bacteria (*Bacillus* sp.) increased with an increase in soil temperature with maximum reduction occurring at 30°C. Bacterial growth and Cr(VI) reduction by the strain *Amphibacillus* sp. KSUCr3 were studied at various temperatures (25–45°C) by Ibrahim et al. (2011). Chromate reduction was increased with temperature up to 40°C, which appeared to be the optimal temperature for growth of the strain KSUCr3. However, they noticed that at 45°C the bacterial growth and Cr(VI) reduction were dramatically decreased. It has been reported that the optimal temperature for reducing Cr(VI) is in the range of 30–37°C (Cheung and Gu 2007). Maximum Cr(VI) reduction occurred at 35°C for *Ochrobactrum* sp. CSCr-3 (He et al. 2009) and *Nesterenkonia* sp. strain MF2 (Amoozegar et al. 2007), but the value was 37°C for *Streptomyces* sp. MS-2 (Mabrouk 2008) and *Rhizopus oryzae* (Sukumar 2010). Biosorption with *Phanerochaete chrysosporium* MTCC 787 removed 99.7% of Cr(VI) at an optimum temperature of 40°C (Pal and Vimala 2011). Chromate reductase from thermophilic *Thermus scotoductus* SA-01 has been recently identified to have an optimum Cr(VI) reduction at a temperature of 65°C (Opperman et al. 2008). In another study, Horton et al. (2006) evaluated the possibility of microbial remediation of Cr(VI) contamination using microorganisms adapted to low temperatures (≤15°C) in aquifers. They identified that *Arthrobacter aurescens* had the potential to reduce Cr(VI) at a low temperature (viz., 10°C).

Qureshi et al. (2003) investigated the temperature and microbial effects on trace element leaching, including As from metalliferous peats. The general ranking of percent losses to leachate and microbial respiration observed among temperature treatments was 28 > 16 > 4 > 37°C. Maximum As recoveries in the leachate were 21.5% for the acidic peat and 5.6% for the neutral peat. The lower microbial activity at 37°C was attributable to fewer bacterial species present that were adapted to that temperature. The effect of temperature on leaching loss of As was attributed to the microbially induced transformation of As. Kim (2010) studied the effect of temperature ($11 \pm 1°C$ and $28 \pm 1°C$) on the adsorption of As(V) onto soil. The results indicated that the highest adsorption was observed at 28°C, implying that the adsorption process was endothermic. Weber et al. (2010) monitored the formation of Fe(II) and As(III) in the pore water and in the soil solid phase during flooding of a contaminated floodplain soil at temperatures of 23, 14, and 5°C. At all temperatures, flooding induced the development of anoxic conditions and caused increasing concentrations of dissolved Fe(II) and As(III). Decreasing the temperature from 23 to 14 and 5°C slowed down soil reduction and Fe and As release.

4.6 Rhizosphere Effects

The rhizosphere influences microbial transformation of trace elements through its effect on microbial activity, pH, and the release of organic compounds. For example, Zayed and Terry (1994) showed that the addition of bacterial antibiotics to nutrient solution substantially inhibited the volatilization of Se, which was linked to the action of antibiotics on bacteria present in the rhizosphere or in the root itself. In vitro Se volatilization, using samples taken from a constructed wetland contaminated with Se(IV), showed that microbial cultures prepared from rhizosphere soils had higher rates of volatilization than cultures prepared from bulk soil (Azaizeh et al. 1997). Most of the Se volatilized by rhizosphere microbes was from bacteria rather than fungi in the sediments. Selenium volatilization by these bacterial cultures was greatly enhanced by aeration and the addition of an energy source, which in in situ, is probably derived from wetland plants.

Mercury methylation is promoted in rhizospheres where microorganisms are dense and organic matter content is high. The rate of Hg methylation is found to be higher in the rhizosphere than non-rhizosphere bulk soil (Sun et al. 2011). Achá et al. (2005) showed that Hg methylation was influenced by rhizospheres of *Polygonum densiflorum* and *Eichhornia crassipes*. Sulfate reducing bacteria activity and Hg methylation potentials were higher in the rhizosphere of *P. densiflorum* compared to that of *E. Crassipes*, which was attributed to the higher C and N concentrations in the former plant species. Carbon content also was associated with sulfate-reducing bacteria and inhibition of sulfate-reducing bacteria resulted in the reduction of Hg methylation.

Plant roots enhance the reduction of metal(loid)s such as As and Cr externally, by releasing root exudates, or internally through endogenous metal(loid) reductase enzyme activity in the root mainly from an increase in microbial activity (Dhankher et al. 2006; Xu et al. 2007). For example, Bolan et al. (2012) observed higher rates of As(V) and Cr(VI) reduction in rhizosphere than non-rhizosphere soils (Fig. 5), which they attributed to several reasons that included increased microbial activity, DOC, and organic acid production, and to decreased pH and Eh in the rhizosphere

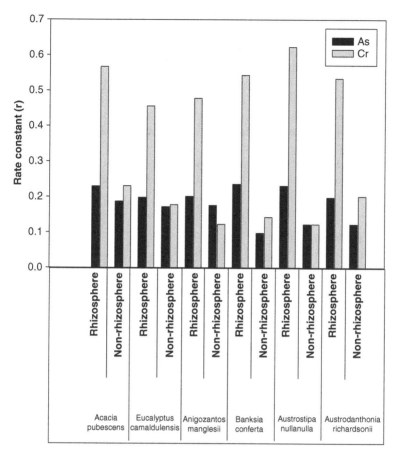

Fig. 5 Rate of reduction of arsenate (As(V)) and chromate (Cr(VI)) in rhizosphere and non-rhizosphere soils of different plant species (Bolan et al. 2012)

soil (Hinsinger et al. 2009). Xu et al. (2007) demonstrated for the first time that tomato and rice roots rapidly reduced As(V) to As(III) internally, some of which was actively effluxed to the growing medium, resulting in an increase in As(III) in the medium.

The pH in the rhizosphere is different than in bulk soil because of the N supply, nutritional status of plants, excretion of organic acids, and CO_2 generation. It is well known that As speciation depends on soil pH. Acidification of the rhizosphere immobilizes As(V) in soil under toxic conditions. Normally the number of microorganisms in the rhizosphere is one order of magnitude greater than in bulk soil, because of higher nutrients and C availability in the rhizosphere (Marschner 1995;

Fitz and Wenzel 2002). Since various bacteria, fungi, yeasts, and algae transform As compounds by oxidation, reduction, demethylation, and methylation, the rhizosphere is an important factor affecting the microbial transformation of As (Frankenberger and Arshad 2002).

The rhizosphere environment can reduce Cr(VI) to Cr(III), because the pH dependent-redox reaction is related to Fe^{2+}, organic matter, and S contents in the rhizosphere (Zeng et al. 2008). Chen et al. (2000) reported that Cr(VI) reduction in a fresh wheat rhizosphere was induced by the decrease of pH. Microbial metabolism in the rhizosphere also caused the reduction of Cr(VI) to Cr(III). Low-molecular-weight organic acids, such as formic and acetic acids in the rhizosphere, can contribute to Cr(VI) reduction and to Cr(III) chelation (Bluskov et al. 2005).

4.7 Soil Amendments

Microbial transformation of trace elements in soils and sediments is affected by inorganic and organic amendments (Park et al. 2011a) (Table 7). Most microbial transformation reactions require an energy source, which is often organic, but can be inorganic too.

Nitrate addition has often been shown to affect the microbial reduction of Se. In general, NO_3^- is a potential inhibitor of Se(VI) reduction under anaerobic condition because it can act as an electron acceptor. Losi and Frankenberger (1997a) observed a correlation between potential dissimilatory Se reduction (DSeR) and potential denitrification, indicating possible involvement of NO_3^--respiring microbes in DSeR. Nitrate or nitrite (NO_2^-) additions cause substantial inhibition of DSeR in drainage sediments and freshwaters. The complete reduction of N oxyanions preceded the reduction of Se(VI); microbial growth was slower on Se(VI) than on NO_3^- and could not be explained on the basis of Eh, which were nearly equivalent (ca. 400 mV) for the two oxyanions.

However, Rech and Macy (1992) observed that *T. selenatis* possessed a different terminal reductase for each of these two electron acceptors (NO_3^- and Se(VI))—a selenate reductase that catalyzes the reduction of Se(VI) to Se(IV) and NO_3^- reductase (NR) that catalyzes the reduction of NO_3^- to NO_2^-. Most other isolates produced a single reductase (NR), which serves as the terminal enzyme for reduction of both NO_3^- and Se(VI). In these cases, when both NO_3^- and Se(VI) are present, only NO_3^- can be used as the terminal electron acceptor; Se(VI) can serve as an electron acceptor only when NO_3^- is absent or is exhausted through initial reduction process.

For *T. selenatis*, the presence of NO_3^- did not affect the conversion of Se(VI) to Se(IV); in fact, NO_3^- was necessary for the further conversion of Se(IV) to Se(0). In cultures grown with NO_3^- plus Se(VI), these electron acceptors were reduced concomitantly, and Se(IV) reduction occurred only in the presence of NO_2^-. It is concluded that NO_2^- reductase (NIR) catalyzed the reduction of both NO_2^- and Se(IV). Nitrite reductase or a component of NIR respiratory system of *T. selenatis* is involved in catalyzing the reduction of Se(IV) to Se(0), while also reducing NO_2^-

Table 7 Selected references on the remediation of arsenic, chromium, mercury, and selenium toxicity by organic amendments

Trace element	Amendment	Plant used	Observation	Reference
Arsenic	Municipal solid waste, biosolids compost	*Pteris vittata*	Increased soil water soluble As and reduced As(V) to As(III), but decreased leaching in the presence of fern	Cao et al. (2003)
	Biosolids compost	*Daucus carota* L and *Lactuca sativa* L	Biosolids compost reduced plant As uptake by 79–86% which might be due to adsorption of As by biosolids organic matter	Cao and Ma (2004)
	Green waste compost and biochar	*Miscanthus* species	Green waste compost substantially increased the plant yield, however increased water soluble and surface adsorbed fractions of As	Hartley et al. (2009)
	Dairy sludge	*Jatropha curcas* L	Application of dairy sludge decreased DTPA-extractable As in soil	Yadav et al. (2009)
Chromium	Cow manure	*Festuca arundinacea*	Reduced Cr in roots and no change in Cr concentration in shoots	Banks et al. (2006)
	Biosolids compost	*Brassica juncea*	Reduced Cr concentration in plant tissue	Bolan et al. (2003a)
	Cattle compost and straw	*Lactuca sativa*	Cr content in aerial biomass decreased with the addition of amendments, which may be due to decreased association of Cr with carbonates and amorphous oxides and increased association with humic substances	Rendina et al. (2006)
	Hog manure and cattle dung compost	*Triticum vulgare*	Hog manure decreased soil available Cr(VI), attributed to its low C/N ratio and thus increased microbial reduction of Cr(VI)	Lee et al. (2006)
	Bark of *Pinus radiata*	*Helianthus annuus*	Reduced availability of Cr for plant uptake	Bolan and Thiagarajan (2001)
	Biosolid compost	*Sesbania punicea* and *Sesbania virgata*	Decreased Cr in plant extract	Branzini and Zubillaga (2010)
	Dairy sludge	*Jatropha curcas* L	Application of dairy sludge decreased DTPA-extractable Cr in soil	Yadav et al. (2009)
	Farm yard manure	*Spinacea oleracea*	Increased root and shoot growth by decreasing Cr(VI) toxicity	Singh et al. (2007)

(continued)

Table 7 (continued)

Trace element	Amendment	Plant used	Observation	Reference
Mercury	Humic acid	*Lactuca sativa* and *Brassica chinensis*	Humic acids decreased the amount of Hg in soil and translocation of Hg into plants	Wang et al. (1997)
	Green waste compost	*Vulpia myuros* L	Addition of compost showed negative relationship with soluble Hg and Hg tissue concentration, which may be due to the adsorption by compost	Heeraman et al. (2001)
	Reactivated carbon	–	Powder reactivated carbon (PAC) increased the stabilization/solidification of Hg in the solid wastes and pretreating the PAC with carbon disulfide (CS_2) increased adsorption efficiency	Zhang and Bishop (2002)
	Fulvic acid	–	Presence of fulvic acid increased the adsorption of Hg on goethite, which might be due to the strong affinity between sulfur groups within the fulvic acid and Hg	Bäckström et al. (2003)
	Humic acid	*Brassica juncea*	Mercury translocation to aerial tissues of plant was restricted in the presence of humic acid	Moreno et al. (2005a)
	thiosulfates	*Brassica juncea*	Thiosulfates increased Hg accumulation in the plant, and Hg could be removed by phytoextraction from contaminated soils	Moreno et al. (2005b)
Selenium	Insoluble (casein) and soluble (casamino acids) organic amendments	–	Organic amendments enhanced Se removal by providing an energy source and methyl donor to the methylating microorganisms, which increased Se volatilization from soil	Zhang and Frankenberger (1999)
	Orange peel, cattle manure, gluten and casein	–	The addition of organic amendments promoted the volatilization of Se; gluten was more effective, increased volatilization by 1.2- to 3.2-fold over the control	Calderone et al. (1990)
	Compost manure and gluten	–	Reduction of Se(VI) to Se(IV) increased in the presence of organic amendments under low oxygen concentration, thereby retarding Se mobility	Guo et al. (1999)
	Press mud and poultry manure	*Triticum aestivum* L and *Brassica napus*	Application of amendments reduced Se accumulation by enhancing volatilization, thereby reducing the transfer of Se from soil to plant	Dhillon et al. (2010)
	Poultry manure, sugar cane press mud, and farmyard manure	*Triticum aestivum* L and *Brassica napus*	Addition of organic amendments decreased Se accumulation and increased grain quality; however, the extent of reduction depended on the type of organic amendment applied	Sharma et al. (2011)

Fig. 6 Effect of organic amendments on the reduction of chromate (Cr(VI)) in soil (Bolan et al. 2003a)

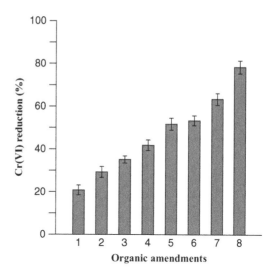

(Rech and Macy 1992). A number of other denitrifying bacteria are also able to reduce Se(IV) under denitrifying conditions, although not as effectively as *T. selenatis*. Selenite reduction is dependent upon the presence of NO_2^-, formed from NO_3^- reduction, and an active NIR.

It has often been shown that addition of organic matter-rich soil amendments enhances the reduction of metal(loid)s such as Cr and Se (Frankenberger and Karlson 1994b; Park et al. 2011a; Fig. 6). For example, several studies showed that the addition of cattle manure enhanced the reduction of Cr(VI) to Cr(III) (Losi et al. 1994; Cifuentes et al. 1996; Higgins et al. 1998). Similarly, various organic material such as powdered leaves, *Pinus sylvestris* bark, black carbon prepared from the dry biomass of rice straw and the plant weed *Solanum elaeagnifolium*, is highly effective in reducing Cr(VI) toxicity (Suseela et al. 1987; Alves et al. 1993; Hsu et al. 2009; Choppala et al. 2012). Soybean meal and rice bran had also been shown to decrease Cr(VI) concentration in soils, which was attributed to Cr(VI) reduction resulting from an increase in DOC content (Chiu et al. 2009). Biomasses of brown seaweed (*Eclonia* sp.), green seaweed (*Ulva lactuca*), red seaweed (*Palmaria palmata* and *Polysiphonia lanosa*), yeast (*Saccharomyces cerevisiae*), and algae (*Sargassum* sp.) have been used to remove Cr(VI) from industrial effluents (Park et al. 2004; Murphy et al. 2008; Parvathi and Nagendran 2008; Vieira et al. 2008). Various reasons could be given for the increase in the reduction of Cr(VI) in the presence of the organic manure composts. These include the supply of C and protons, and the stimulation of microorganisms that are considered to be the major factors enhancing the reduction of Cr(VI) to Cr(III) (Losi et al. 1994). For example, Choppala (2011) observed a decrease in Cr(VI) toxicity in soils treated with black carbon, which was attributed to the supply of electrons for the reduction of toxic Cr(VI) to nontoxic Cr(III) species (Plate 1).

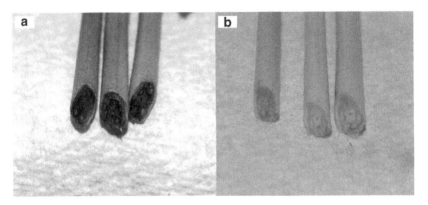

Plate 1 Effect of black carbon on chromate (Cr(VI)) phytotoxicity in contaminated soils (**a**) stems of plants grown in Cr(VI) contaminated soils, (**b**) stems of plants grown black carbon amended Cr(VI) contaminated soils (Choppala 2011)

With the same amount of total organic C addition, Bolan et al. (2003a) and Choppala et al. (2012) observed significant difference in the extent of Cr(VI) reduction among various organic manure composts (Fig. 6). The extent of Cr(VI) reduction increased with increasing level of DOC added through manure addition, which has been identified to facilitate the reduction of Cr(VI) to Cr(III) in soils (Jardine et al. 1999; Nakayasu et al. 1999; Chiu et al. 2009; Hsu et al. 2010). For example, the hydroquinone groups in organic matter have been identified as the major electron donor source for reducing Cr(VI) to Cr(III) in soils (Choppala et al. 2012).

The easily oxidizable organic C fractions, such as DOC, provide the energy source for the soil microorganisms involved in the reduction of metal(loid)s [e.g., Cr(VI)] and nonmetal(loid)s (e.g., NO_3^-) (Paul and Beauchamp 1989; Jardine et al. 1999; Vera et al. 2001; Bolan et al. 2011). Although manure addition induces the remediation of Cr-contaminated soils by reducing mobile toxic Cr(VI) to nontoxic and less mobile Cr(III), it is likely that microbial transformations of the aromatic As and Hg compounds could occur, which in turn, may result in the production of more toxic inorganic species (Cullen and Reimer 1989; Kumagai and Sumi 2007).

Chromate can be biotransformed, even in the presence of different electron acceptors that include O_2, NO_3, sulfate, and Fe. Sugar addition had the greatest effect on enhancing Cr(VI) removal (Tseng and Bielefeldt 2002). Tseng and Bielefeldt (2002) noticed that less DOC was consumed per unit amount of Cr(VI) transformed under anaerobic conditions [0.8–93 mg DOC mg Cr(VI)$^{-1}$], compared with aerobic conditions [1.4–265 mg DOC mg Cr(VI)$^{-1}$].

Oliver et al. (2003) reported from using both the batch and column experiments that Cr(VI) reduction and loss from the aqueous phase were enhanced by adding high levels of both NO_3^- and organic C (molasses). Nutrient amendments resulted in up to 87% reduction of the initial 67 mg L^{-1} Cr(VI) in an unsaturated

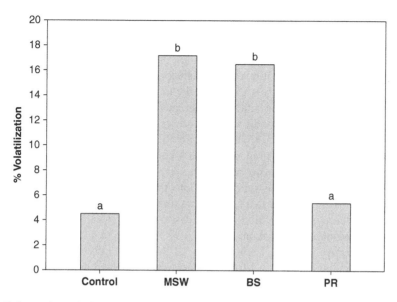

Fig. 7 Loss of arsenic in chromated copper–arsenate soil (*MSW* municipal solid waste, *BS* biosolid, *PR* phosphate rock) (Cao et al. 2003)

batch experiment. Molasses and NO_3^- additions to unsaturated flow columns receiving 65 mg L^{-1} Cr(VI) enhanced microbially mediated reduction and immobilization of Cr.

The addition of Fe(II) decreased net Hg methylation in sediments and sulfide (S(−II)) concentration. The reduction in net Hg methylation can be attributed to the decreased concentration of uncharged, bioavailable Hg, which is positively related to S(−II) (Mehrotra and Sedlak 2005). The effect of clay addition on the Hg methylation depends on surface coatings. However, clays prevent Hg methylation through adsorption or sometimes promote demethylation by microorganisms. Humic substance facilitates Hg methylation, while humic coatings on clay stimulated demethylation in freshwater sediments (Jackson 1989; Zhang and Hsu-Kim 2010).

Most As forms As sulfides, such as realgar (AsS), orpiment (As_2S_3), and arsenopyrite (FeAsS). These have low solubility and mobility when S is abundant, as a result of biosolid amendments in soil. Organic amendments such as biosolids and manure significantly reduce the potential environmental risks of As contamination under highly anoxic conditions (Carbonell-Barrachina et al. 1999). Similarly, Yadav et al. (2009) showed that the addition of dairy sludge and biofertilizer reduced the bioavailability of As and Cr, and promoted plant growth. The addition of manure increased the loss of As from contaminated soil through methylation to volatile As by microbes (Fig. 7). The rates of As loss was closely related to the microbial respiration because of nutrient supplementation to microbes. Bioaugmentation by As-methylating fungi increased H_2As evolution rates in field-contaminated soils (Edvantoro et al. 2004).

5   Implications for Bioavailability and Remediation

Bioaccumulation and microbial transformation processes play a major role in the decontamination and remediation of trace element contaminated soil, sediment, and water (Fig. 1). Microorganisms play a vital role in transforming trace elements, thereby influencing their bioavailability and remediation (Fig. 8). From toxicological or environmental viewpoints, these processes are important for three reasons. They may alter (a) the toxicity, (b) the water solubility, and/or (c) the mobility of the element (Alexander 1999). An increase in solubility and mobility can be exploited to bioremediate insoluble forms of elements in soil, because the biotransformed product is released from the solid phase into the solution phase. Conversely, a decrease in element solubility can be used to remove the element from surface or groundwater through precipitation. In some cases, gaseous metal(loid) products can be removed through volatilization.

The biosorptive process of removing metal(loid)s generally lack specificity in metal(loid) binding and is sensitive to ambient environmental conditions, such as pH, solution composition, and the presence of metal(loid) chelators. Genetically engineered microorganisms (e.g., *E. coli*), which express metal(loid) binding protein (i.e., metallothionein) and a metal(loid)-specific transport system, are selective for accumulation of specific metal(loid)s in the presence of high concentrations of other metal(loid)s and chelating agents in solution (Chen and Wilson 1997; Krämer et al. 2007). These organisms can be used to remove specific metal(loid)s from contaminated soil and sediments by washing these matrices with chelating agents

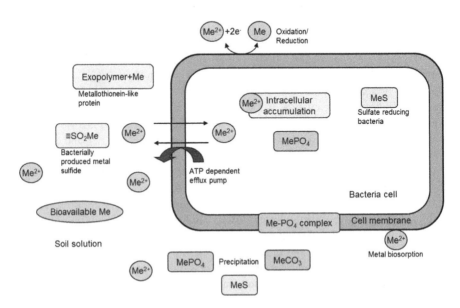

Fig. 8 Microbial aspects of trace element immobilization (Park et al. 2011c)

and then regenerating the chelating agents in a reactor containing the microbial strain (Bolan et al. 2006; Perpetuo et al. 2011).

Incorporating the genetically engineered metalloregulatory protein onto the surface of *E. coli* cells showed high affinity and selectivity for Hg sorption. The presence of surface-exposed metalloregulatory protein increased Hg biosorption by sixfold, compared to the wild-type cells, which could be applied for the cleanup of other metal(loid)s (Bae et al. 2003). Murugesan et al. (2006) showed that the tea fungus produced during black tea fermentation removed 100% of As(III) and 77% of As(V) in groundwater.

Dissimilatory metal(loid) reduction has the potential to be helpful for both intrinsic and engineered bioremediation of contaminated environments. Dissimilatory reduction of Se, Cr, and possibly other metal(loid)s can convert soluble metal(loid) species to insoluble forms that can readily be removed from contaminated waters or waste streams (Crowley and Dungan 2002). Reduction of Hg can volatilize Hg from surface water and ocean (Lovley 1995; Moreno et al. 2005a, b). Arsenic can be reduced to As(0), which is subsequently precipitated as As_2S_3 as a result of microbial sulfate reduction. Because As(III) is more soluble than As(V), the latter can be reduced using bacteria in soil and subsequently leached. Conversely, As(III) is oxidized to As(V) using microbes, which is subsequently precipitated using ferric ions (Williams and Silver 1984). *Desulfotomaculum auripigmentum* reduces both As(V) to As(III) and sulfate to H_2S, and leads to As_2S_3 precipitation (Newman et al. 1997).

Because of the lower solubility of Cr(III) than Cr(VI), the reduction reaction will eventually result in the immobilization of Cr, thereby diminishing the mobility and transport. Reduction of Cr(VI) to Cr(III) and subsequent hydroxide precipitation of Cr(III) ion is the most common method for treating Cr(VI)-contaminated industrial effluents (Blowes et al. 1997; James 2001). Similarly, Choppala (2011) has noticed that reducing Cr(VI) to Cr(III) in variable charge soils is likely to result in the adsorption of Cr(III) (4) from an increase in pH-induced negative charges and precipitation of $Cr(OH)_3$ (5), resulting from the reduction-induced release of OH⁻ ions (Fig. 9). Microbial reduction can be accomplished by direct reduction of Cr(VI) to Cr(III), using C as an energy source. Indirect reduction is achieved by sulfate addition under reduced conditions resulting in the release of H_2S which subsequently reduces Cr(VI) to Cr(III), and causes the precipitation of Cr as Cr_2S_3 (Smith et al. 2002; Vainshtein et al. 2003).

$$Cr(III) + Soil \rightarrow Cr^{3+} + H_2O + H^+ \left(adsorption - proton\ release\ reaction\right). \quad (4)$$

$$Cr(III) + H_2O \rightarrow Cr(OH)_3 + H^+ \left(precipitation - proton\ release\ reaction\right). \quad (5)$$

Selenium is subject to several forms of microbial transformations, some of which may have applicability in bioremediation (Ehrlich 1996; Losi and Frankenberger 1997b). Biological immobilization of Se(VI) by dissimilatory reduction to Se(0) is a practical approach for remediation. Anaerobic bacteria can be grown in the contaminated medium with a C source, such as acetate as an electron donor and Se(VI) an electron acceptor [(6) and (7)]. The extent of Se(VI) reduction depends upon the

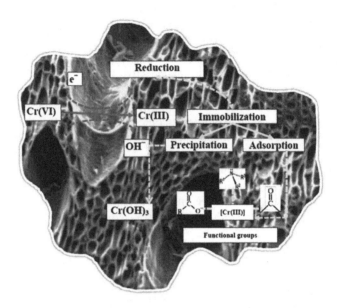

Fig. 9 Concomitant reduction and immobilization of chromium in black carbon amended soils (Choppala 2011)

availability of the C source, and the reduced Se in the elemental form, which is insoluble can be physically separated from contaminated water. Similarly, the methylation reaction can be used to form gaseous metal(loid) species, which can easily be removed through volatilization (Thompson-Eagle and Franakenberger 1992; Bañuelos and Li 2007).

$$4CH_3COO^- + 3SeO_4^{2-} \rightarrow 3Se^0 + 8CO_2 + 4H_2O + 4H^+. \tag{6}$$

$$CH_3COO^- + 4SeO_4^{2-} \rightarrow 4SeO_3^{2-} + 2HCO_3^- + H^+. \tag{7}$$

Bioremediation of metal(loid) contaminated sites can also be achieved indirectly through the transformation of other compounds. For example, bioleaching of metal(loid)s using acidification produced during the biological oxidation of pyrite has practical application for removing metal(loid)s in tailings and coal mine spoils (Sand et al. 2001). Similarly, microorganisms produce H_2S from sulfate and inorganic P from organic P; both H_2S and inorganic P form insoluble precipitates with a number of metal(loid)s (Douglas and Beveridge 1998; Bolan et al. 2003b; Kosolapov et al. 2004). Park et al. (2011b) noted that the inoculation of P-solubilizing bacteria enhanced the immobilization of Pb, thereby reducing its subsequent mobility and bioavailability.

Bender et al. (1995) documented the commercial application of microbial mats constructed by combining cyanobacteria inoculum with a sediment inoculum from a metal(loid)-contaminated site for removing and transforming metal(loid)s. The mat, comprised of a heterotrophic and autotrophic community, tolerated high concentrations of toxic metal(loid)s, such as Cd, Pb, Cr, Se, and As (up to 350 mg L^{-1}), and achieved a rapid removal of the metal(loid)s from contaminated water. The toxic metal(loid)s were managed by the mat because of (1) the production of metal(loid)-binding polyanionic polysaccharides; (2) deposition of metal(loid) compounds outside the cell surfaces; and (3) chemical modification of the aqueous environment surrounding the mat. Thus, depending on the chemical character of the microzone of the mat, the sequestered metal(loid)s can be oxidized, reduced, and precipitated as sulfides or oxides. Precipitates of red amorphous Se(0), for instance, were identified in the mat exposed to Se(VI).

Roeselers et al. (2008) reviewed the actual and potential applications of microbial mats in wastewater treatment, bioremediation, fish-feed production, biohydrogen production, and soil improvement. They suggested that phototrophic biofilms would also be suitable for developing inexpensive treatment methods for developing countries, where land values are relatively low and domestic and industrial wastewaters are still discharged untreated. Microbial mats offer several advantages that include low cost, durability, ability to function in both fresh and salt water, tolerance to high concentrations of metal(loid)s, and the unique capacity of mats to form associations with new microbial species.

Industrial fermentations discard waste *Bacillus* sp. that can be treated with alkali to increase heavy metal(loid) intake. Granulation of these waste *Bacillus* sp. increases cross-links and retention of Cr and Hg up to 10% of the dry weight, with removal efficiency of >99%, in the concentration range of 10–50 μg L^{-1}. The contaminated metal(loid)s that are loaded onto the biomass granules can be recovered by washing the biomass with sulfuric acid/sodium hydroxide or other complexing agents (White et al. 1995). Heavy metal(loid) contaminated hazardous soils can be detoxified through bacterially mediated reduction. For example, applying black carbon or biochar in the Cr(VI)-contaminated soils increases microbial activity, thereby enhancing Cr(VI) reduction (Choppala et al. 2012). External coating of the copper–chromate–arsenic (CCA)-treated timber posts with black carbon is one of the viable options to arrest Cr(VI) leaching to groundwater.

6 Conclusions and Future Research Needs

Although there have been increasing efforts to reduce the input of trace elements to the biosphere, their introduction to the environment has increased. The relationship between environmental trace-element concentration and their bioavailability appears to be nonlinear. For example, the presence of phosphates, iron, and aluminum compounds in a typical municipal sewage sludge enhances the retention of metal(loid)s, thereby inducing the "plateau effect" in their uptake by crops and preventing the increased metal(loid) bioavailability suggested in the "time bomb" hypothesis (Brown et al. 1998). Nevertheless, bioavailability of certain metal(loid)s continue to increase despite a decrease in their total input to the environment. For example, in Scandinavia, Hg levels in freshwater fish are increasing even though emissions and deposition of Hg have been decreasing for decades (de Lacerda and Salomons 1998; Danielsson et al. 2011). The contamination of fish has been attributed largely to microbial remobilization of Hg that was locked in watershed soils and sediments for decades (Salomons and Stigliani 1995; Wang et al. 2012). Similarly, exploitation of groundwater for drinking and irrigation purposes has altered the physicochemical environment of vadose zone releasing large quantities of As in Indo-Gangetic region (Bhattacharya et al. 1997; Marechal et al. 2006).

Two important issues need to be addressed when applying biotransformation processes to manage trace elements contamination. Firstly, implementation of bioremediation methods should be done with caution because many sites contain multiple metal(loid)s, organic compounds, and organisms that affect the output of bioremediation approaches. Therefore, remediation of contaminated sites usually requires a combination of many different approaches. Rapid developments in the application of synchrotron-based techniques have generated new methods capable of characterizing the distribution, and speciation of major and trace elements at the surfaces of geological and biological materials. These molecular tools can also detect dynamic chemical transformations that occur between microbial biofilms and the adjacent minerals, which enable the study of integrated systems. For example, when profiling the partitioning of Se(VI) within *Burkholderia cepacia* biofilms, Templeton et al. (2003) unexpectedly noticed that *B. cepacia* rapidly reduced Se(VI) to Se(IV) and Se(0), and the Se(IV) intermediate diffuses through the biofilm to preferentially bind to the underlying mineral surface. The rapid stratification of Se redox-species would not have been observed, nor predicted, using bulk spectroscopic techniques or studying biofilms or mineral surfaces alone.

Secondly, bioremediation rarely restores an environment to its original condition. Often, the residual contamination left after treatment is strongly sorbed and not available to microorganisms for degradation. Over a long period of time, these residuals can be slowly released, generating additional pollution. There is little research on the fate and potential toxicity of such released residuals; therefore, both the public and regulatory agencies continue to be concerned about the possible deleterious effects of residual contamination. Future areas of research to better understand bioavailability and bioaccumulation of metal(loid)s involving soil microorganisms include:

- The realization that microbes operate as consortia of organisms rather than as single cells requires attention to be focused on the microenvironments in which chemical transformations occur through microbial consortia.
- Development and validation of empirical and mechanistic kinetic models, linking soil physicochemical characteristics and metal(loid) speciation with bioaccumulation and microbial transformation.
- For Se, better tools for the *in silico* identification of selenoproteins and further investigation on Se respiration are needed.
- For As, further investigation is needed on how the organoarsenicals are transformed, and on the processes of methylation and demethylation.

- For Cr, the characterization of C, which includes aliphatic and aromatic C in DOC, needs to be quantified. Moreover, an in-depth study on the active role that electrons and protons play on Cr(VI) reduction is needed using synchrotron-based studies.
- For Hg, the pathways for methyl Hg degradation in the environment are only partially understood. There is an urgent need to identify pure cultures that degrade methyl Hg oxidatively so that research on the biochemistry of this process can commence.
- There is a need to understand the link between the enormous diversity of Hg-transforming organisms, which is revealed by molecular genetic studies (e.g., Liebert et al. 1997; Bogdanova et al. 2001), and their interactions with Hg so that the management and remediation of Hg contamination under diverse conditions could be enhanced.
- Efforts must be placed in determining the environmental factors (biotic and abiotic) enhancing methyl Hg production, and also in identifying and characterizing the genetic/enzymatic systems, controlling Hg methylation.

7 Summary

The dynamics of trace elements in soils depend on both their physicochemical interactions with inorganic and organic soil constituents and biological interactions associated with the microbial activities of soil–plant systems. Microorganisms control the transformation (microbial or biotransformation) of trace elements by various mechanisms that include oxidation, reduction, methylation, demethylation, complex formation, and biosorption. Microbial transformation processes can influence the solubility and subsequent mobility of these trace elements, especially As, Cr, Hg, and Se in soils and sediments by altering their speciation and redox state. These processes play a major role in the bioavailability, mobility, ecotoxicology, and environmental health of these trace elements. Microbial transformation processes can be readily managed and enhanced for efficient removal of contaminants, provided the biochemistry of these processes is understood. Thus, a greater understanding of microbial transformation processes will help to monitor the environmental fate of the trace elements, particularly through the food web, and will help to develop in situ bioremediation technologies. In this review the key microbial transformation processes, including biosorption, redox reactions, methylation/demethylation reactions controlling the fate and behavior of As, Cr, Hg, and Se are addressed. The factors affecting these processes in relation to the bioavailability and remediation of trace elements in the environment are also examined and plausible future research directions are suggested.

The microbial transformation processes can be viewed as protective mechanisms available to microorganisms that increase their resistance to toxic trace elements. These mechanisms are biochemical in nature, and generally render the metal(loid) ion ineffective in disturbing the normal biochemical processes of the cell. The mechanisms include: efflux pumps that remove the ions from the cell; enzymatic reduction of metal(loid)s to less toxic elemental forms; chelation by enzymatic polymers (i.e., metallothionein); binding the metal(loid) to cell wall surfaces; precipitating insoluble inorganic complexes (usually sulfides and oxides) at the cell surface; biomethylation with subsequent transport through the cell membrane by diffusion. Soil amendments can be used to manipulate microbial transformation processes, thereby managing the bioavailability and remediation of trace elements.

Because many of the environments that receive trace element-containing wastes can be characterized as anoxic, for example, subsurface saturated soils or organic rich marsh sediments, biochemically mediated transformations of metal(loid)s play a vital role in their mobilization and bioavailability. Desorption and remobilization of metal(loid)s, such as Cr and As from sediments, are controlled by pH, Eh, and metal(loid) concentration in the sediment interstitial water, as well as by contents in total Fe, Mn, and mineral hydrous oxides. Physical disturbances of the sediments by storm or flooding may move the underlying sediments to oxidizing environments where the sulfides undergo oxidation resulting in the release of large quantities of metal(loid)s into the water. Chemolithotropic bacteria play a major role in the oxidation process, thereby enhancing the mobilization of metal(loid)s. The use of fungi species offers a way to leach As from industrial waste sites. The production of organic acids with the use of heterotrophic organisms, and the generation of sulfuric acid with the use of microorganism, such as *Thiobacillus*, also offers some promising approaches to the extraction of As. Similarly, depending on the nature of metal(loid)s present in soil, the rhizosphere-induced redox reactions have implications to both their bioavailability to higher plants and microorganisms, and remediation of contaminated soils. For example, while rhizoreduction decreases Cr bioavailability it increases that of As.

While methylation of inorganic Hg and Se in aquatic systems is the norm, it can also occur in the terrestrial environment. As Hg(0) and Se(0) formation removes reactive Hg and Se from soil, sediments and water where these could otherwise be methylated, methylation process plays an important role in their eventual removal from these systems. Despite its potential, there has yet been limited applied research into the use of dissimilatory metal(loid) reduction as a bioremediation tool. Most bioremediation technologies are based on microbial transformation processes that are designed to remove metal(loid)s mainly from aquatic systems. The viability and metabolic activity of microorganisms are the major limiting factors on efficiency of biotransforming metal(loid)s in soils. Even though genetically engineered microorganisms showed high microbial transformation of metal(loid)s, the application of genetically modified organisms into the environment is a matter of concern for many. Therefore it is important to manipulate these microbial transformation reactions by controlling the factors affecting them and also by using appropriate soil amendments. This will enable the sustainable management of contamination by trace elements to mitigate their environmental and health impacts.

Acknowledgments The Postdoctoral fellowship program (PJ008650042012) at National Academy of Agricultural Science, Rural Development Administration, Republic of Korea, supported Dr Kunhikrishnan's contribution.

References

Achá D, Hintelmann H, Yee J (2011) Importance of sulfate reducing bacteria in mercury methylation and demethylation in periphyton from Bolivian Amazon region. Chemosphere 82: 911–916

Achá D, Iñiguez V, Roulet M, Guimarães JRD, Luna R, Alanoca L, Sanchez S (2005) Sulfate-reducing bacteria in floating macrophyte rhizospheres from an Amazonian floodplain lake in Bolivia and their association with Hg methylation. Appl Environ Microbiol 71:7531–7535

Adriano DC (2001) Trace elements in terrestrial environments: biogeochemistry, bioavailability and risks of metals, 2nd edn. Springer, New York

Adriano DC, Wenzel WW, Vangronsveld J, Bolan NS (2004) Role of assisted natural remediation in environmental cleanup. Geoderma 122:121–142

Ahalya N, Ramachandra T, Kanamadi R (2003) Biosorption of heavy metals. Res J Chem Environ 7:71–79

Akagi H, Malm O, Branches FJP, Kinjo Y, Kashima Y, Guimares TRD, Oliveira RB, Haraguchi K, Pfeiffer WC, Takizawa Y, Kato H (1995) Human exposure to mercury due to gold mining in the Tapajos River Basin, Amazon, Brazil: speciation of mercury in human hair, blood and urine. Water Air Soil Pollut 80:85–94

Al Rmalli SW, Dahmani AA, Abuein MM, Gleza AA (2008) Biosorption of mercury from aqueous solutions by powdered leaves of castor tree (*Ricinus communis* L.). J Hazard Mater 152:955–959

Alexander M (1999) Biodegradation and bioremediation, 2nd edn. Academic, San Diego, CA

Alexander M (2000) Aging, bioavailability, and overestimation of risk from environmental pollutants. Environ Sci Technol 34:4259–4265

Allard B, Arsenie I (1991) Abiotic reduction of mercury by humic substances in aquatic system—an important process for the mercury cycle. Water Air Soil Pollut 56:457–464

Alloway B (1990) 2 Soil processes and the behaviour of metals. In: Alloway BJ (ed) Heavy metals in soils. Blackie and Son Ltd., Glasgow

Alves M, Gonzalez BCG, De Carvalho GR, Castenheira JM, Pereira SMC, Vasconcelos LAT (1993) Chromium removal in tannery wastewaters—polishing by *Pinus sylvestris* bark. Water Res 27:1333–1338

Amoozegar MA, Ghasemi A, Razavi MR, Naddaf S (2007) Evaluation of hexavalent chromium reduction by chromate-resistant moderately halophile, *Nesterenkonia* sp. strain MF2. Process Biochem 42:1475–1479

Anand P, Isar J, Saran S, Saxena RK (2006) Bioaccumulation of copper by *Trichoderma viride*. Bioresour Technol 97:1018–1025

Anderson LCD, Bruland KW (1991) Biogeochemistry of arsenic in natural waters: the importance of methylated species. Environ Sci Technol 25:420–427

Aposhian HV, Zakharyan RA, Avram MD, Kopplin MJ, Wollenberg ML (2003) Oxidation and detoxification of trivalent arsenic species. Toxicol Appl Pharmacol 193:1–8

Apte AD, Tare V, Bose P (2006) Extent of oxidation of Cr (III) to Cr (VI) under various conditions pertaining to natural environment. J Hazard Mater 128:164–174

Aryal M, Ziagova M, Liakopoulou-Kyriakides M (2010) Study on arsenic biosorption using Fe (III)-treated biomass of *Staphylococcus xylosus*. Chem Eng J 162:178–185

Azaizeh HA, Gowthaman S, Terry N (1997) Microbial selenium volatilization in rhizosphere and bulk soil from a constructed wetland. J Environ Qual 26:666–672

Bachate SP, Khapare RM, Kodam KM (2012) Oxidation of arsenite by two β-proteobacteria isolated from soil. Appl Microbiol Biotechnol 93:2135–2145

Bäckström M, Dario M, Karlsson S, Allard B (2003) Effects of a fulvic acid on the adsorption of mercury and cadmium on goethite. Sci Total Environ 304:257–268

Bae W, Wu CH, Kostal J, Mulchandani A, Chen W (2003) Enhanced mercury biosorption by bacterial cells with surface-displayed MerR. Appl Environ Microbiol 69:3176–3180

Bahlmann E, Ebinghaus R, Ruck W (2006) Development and application of a laboratory flux measurement system (LFMS) for the investigation of the kinetics of mercury emissions from soils. J Environ Manage 81:114–125

Baker MD, Inniss WE, Mayfield CI (1983) Effect of pH on the methylation of mercury and arsenic by sediment microorganisms. Environ Technol Lett 4:89–100

Banks M, Schwab A, Henderson C (2006) Leaching and reduction of chromium in soil as affected by soil organic content and plants. Chemosphere 62:255–264

Bañuelos GS, Li ZQ (2007) Acceleration of selenium volatilization in seleniferous agricultural drainage sediments amended with methionine and casein. Environ Pollut 150:306–312

Baralkiewicz D, Gramowska H, Gołdyn R, Wasiak W, Kowalczewska-Madura K (2007) Inorganic and methyl-mercury speciation in sediments of the Swarz dzkie Lake. Chem Ecol 23:93–103

Barnhart J (1997) Chromium chemistry and implications for environmental fate and toxicity. Soil Sediment Contam 6:561–568

Battaglia-Brunet F, Dictor MC, Garrido F, Crouzet C, Morin D, Dekeyser K, Clarens M, Baranger P (2002) An arsenic (III) oxidizing bacterial population: selection, characterization, and performance in reactors. J Appl Microbiol 93:656–667

Bender J, Lee RF, Phillips P (1995) A review of the uptake and transformation of metals and metalloids by microbial mats and their use in bioremediation. J Ind Microbiol 14:113–118

Benoit JM, Gilmour CC, Mason RP, Heyes A (1999) Estimation of mercury-sulfide speciation and bioavailability in sediment and porewaters. Environ Toxicol Chem 18:951–957

Berman M, Bartha R (1986) Levels of chemical versus biological methylation of mercury in sediments. Bull Environ Contam Toxicol 36:401–404

Bhandari N, Reeder RJ, Strongin DR (2011) Photoinduced oxidation of arsenite to arsenate on ferrihydrite. Environ Sci Technol 45:2783–2789

Bhattacharya P, Chatterjee D, Jacks G (1997) Occurrence of arsenic-contaminated groundwater in alluvial aquifers from delta plains, Eastern India: options for safe drinking water supply. Int J Water Resour Dev 13:79–92

Bishnoi NR, Kumar R, Kumar S, Rani S (2007) Biosorption of Cr (III) from aqueous solution using algal biomass *Spirogyra* spp. J Hazard Mater 145:142–147

Bisinoti MC, Junior E, Jardim WF (2007) Seasonal behavior of mercury species in waters and sediments from the Negro River Basin, Amazon, Brazil. J Braz Chem Soc 18:544–553

Biswas KC, Barton LL, Tsui WL, Shuman K, Gillespie J, Eze CS (2011) A novel method for the measurement of elemental selenium produced by bacterial reduction of selenite. J Microbiol Methods 86:140–144

Blowes DW, Ptacek CJ, Jambor JL (1997) In-situ remediation of Cr (VI)-contaminated groundwater using permeable reactive walls: laboratory studies. Environ Sci Technol 31:3348–3357

Bluskov S, Arocena J, Omotoso O, Young J (2005) Uptake, distribution, and speciation of chromium in *Brassica juncea*. Int J Phytoremediation 7:153–165

Boening DW (2000) Ecological effects, transport, and fate of mercury: a general review. Chemosphere 40:1335–1351

Bogdanova E, Minakhin L, Bass I, Volodin A, Hobman JL, Nikiforov V (2001) Class II broad-spectrum mercury resistance transposons in gram-positive bacteria from natural environments. Res Microbiol 152:503–514

Bolan NS, Adriano DC, Kunhikrishnan A, James T, McDowell R, Senesi N (2011) Dissolved organic matter: biogeochemistry, dynamics, and environmental significance in soils. Adv Agron 110:1–75

Bolan NS, Adriano DC, Natesan R, Koo BJ (2003a) Effects of organic amendments on the reduction and phytoavailability of chromate in mineral soil. J Environ Qual 32:120–128

Bolan NS, Adriano DC, Naidu R (2003b) Role of phosphorus in (im)mobilization and bioavailability of heavy metals in the soil-plant system. Rev Environ Contam Toxicol 177:1–44

Bolan N, Kunhikrishnan A, Gibbs J (2012) Rhizoreduction of arsenate and chromate in Australian native grass, shrub and tree vegetation. Plant Soil (DOI 10.1007/s11104-012-1506-y)

Bolan NS, Mahimairaja S, Megharaj M, Naidu R, Adriano DC (2006) Biotransformation of arsenic in soil and aquatic environments: bioavailability and bioremediation. In: Naidu R, Smith E, Owens G, Bhattacharya P, Nadebaum P (eds) Managing arsenic in the environment: from soil to human health. CSIRO, Melbourne, pp 433–453

Bolan NS, Thiagarajan S (2001) Retention and plant availability of chromium in soils as affected by lime and organic matter amendments. Aust J Soil Res 39:1091–1104

Boszke L, Kowalski A, Glosinska G, Szarek R, Siepak J (2003) Environmental factors affecting speciation of mercury in the bottom sediments; an overview. Pol J Environ Stud 12:5–14

Bowell R (1994) Sorption of arsenic by iron oxides and oxyhydroxides in soils. Appl Geochem 9:279–286

Branzini A, Zubillaga M (2010) Assessing phytotoxicity of heavy metals in remediated soil. Int J Phytoremediation 12:335–342

Bridou R, Monperrus M, Gonzalez PR, Guyoneaud R, Amouroux D (2011) Simultaneous determination of mercury methylation and demethylation capacities of various sulfate reducing bacteria using species specific isotopic tracers. Environ Toxicol Chem 30:337–344

Brock TD, Madigan MT (1991) Biology of microorganisms. Prentice Hall, Englewood Cliffs, NJ

Brown S, Chaney R, Angle JS, Ryan JA (1998) The phytoavailability of cadmium to lettuce in long-term biosolid amended soil. J Environ Qual 27:1071–1078

Calderone S, Frankenberger W, Parker D, Karlson U (1990) Influence of temperature and organic amendments on the mobilization of selenium in sediments. Soil Biol Biochem 22:615–620

Camargo FA, Okeke BC, Bento FM, Frankenberger WT (2003) In vitro reduction of hexavalent chromium by a cellfree extract of *Bacillus* sp. ES 29 stimulated by Cu^{2+}. Appl Microbiol Biotechnol 62:569–573

Cao X, Ma LQ (2004) Effects of compost and phosphate on plant arsenic accumulation from soils near pressure-treated wood. Environ Pollut 132:435–442

Cao X, Ma LQ, Shiralipour A (2003) Effects of compost and phosphate amendments on arsenic mobility in soils and arsenic uptake by the hyperaccumulator, *Pteris vittata* L. Environ Pollut 126:157–167

Carbonell-Barrachina AA, Jugsujinda A, Sirisukhodom S, Anurakpongsatorn P, Burló F, DeLaune RD, Patrick WH Jr (1999) The influence of redox chemistry and pH on chemically active forms of arsenic in sewage sludge-amended soil. Environ Int 25:613–618

Casiot C, Bruneel O, Personne JC, Leblanc M, Elbaz-Poulichet F (2004) Arsenic oxidation and bioaccumulation by the acidophilic protozoan, *Euglena mutabilis*, in acid mine drainage (Carnoules, France). Sci Total Environ 320:259–267

Chakraborty S, Bardelli F, Charlet L (2010) Reactivities of Fe(II) on calcite: selenium reduction. Environ Sci Technol 44:1288–1294

Chen CP, Juang KW, Lin TH, Lee DY (2010) Assessing the phytotoxicity of chromium in Cr (VI)-spiked soils by Cr speciation using XANES and resin extractable Cr(III) and Cr(VI). Plant Soil 334:299–309

Chen NC, Kanazawa S, Horiguchi T (2000) Chromium(VI) reduction in wheat rhizosphere. Pedosphere 10:31–36

Chen SL, Wilson DB (1997) Genetic engineering of bacteria and their potential for Hg^{2+} bioremediation. Biodegradation 8:97–103

Cheung KH, Gu JD (2007) Mechanism of hexavalent chromium detoxification by microorganisms and bioremediation application potential: a review. Int Biodeter Biodegr 59:8–15

Chiu CC, Cheng CJ, Lin TH, Juang KW, Lee DY (2009) The effectiveness of four organic matter amendments for decreasing resin-extractable Cr(VI) in Cr(VI)-contaminated soils. J Hazard Mater 161:1239–1244

Chiu VQ, Hering JG (2000) Arsenic adsorption and oxidation at manganite surfaces. 1. Method for simultaneous determination of adsorbed and dissolved arsenic species. Environ Sci Technol 34:2029–2034

Choi SC, Bartha R (1994) Environmental factors affecting mercury methylation in estuarine sediments. Bull Environ Contam Toxicol 53:805–812

Choppala G (2011) Reduction and bioavailability of chromium in soils. Doctoral thesis, University of South Australia, Australia

Choppala GK, Bolan NS, Megharaj M, Chen Z, Naidu R (2012) The influence of biochar and black carbon on reduction and bioavailability of chromate in soils. J Environ Qual 41:1175–1184

Christen K (2001) Chickens, manure, and arsenic. Environ Sci Technol 35:184A–185A

Ciesielski T, Pastukhov MV, Szefer P (2010) Bioaccumulation of mercury in the pelagic food chain of the Lake Baikal. Chemosphere 78:1378–1384

Cifuentes F, Lindemann W, Barton L (1996) Chromium sorption and reduction in soil with implications to bioremediation. Soil Sci 161:233

Cossich ES, da Silva EA, Tavares CRG, Filho LC, Ravagnani TMK (2004) Biosorption of chromium (III) by biomass of seaweed *Sargassum* sp. in a fixed-bed column. Adsorption 10:129–138

Costa M, Liss P (1999) Photoreduction of mercury in sea water and its possible implications for Hg^0 air-sea fluxes. Mar Chem 68:87–95

Crowley DE, Dungan RS (2002) Metals: microbial processes affecting metals, Encyclopedia of environmental microbiology. Wiley, New York, pp 1878–1893

Cullen WR, Reimer KJ (1989) Arsenic speciation in the environment. Chem Rev 89:713–729

Danielsson S, Hedman J, Miller A, Bignert A (2011) Mercury in Perch from Norway, Sweden and Finland—geographical patterns and temporal trends. Report nr 8:2011, Department of contaminant research, Swedish museum of natural history, Stockholm, Sweden

Das N, Vimala R, Karthika P (2008) Biosorption of heavy metals—an overview. Ind J Biotechnol 7:159–169

Das SK, Guha AK (2009) Biosorption of hexavalent chromium by *Termitomyces clypeatus* biomass: kinetics and transmission electron microscopic study. J Hazard Mater 167:685–691

de Lacerda L (2003) Updating global Hg emissions from small-scale gold mining and assessing its environmental impacts. Environ Geol 43:308–314

de Lacerda LD, Salomons W (1998) Mercury from gold and silver mining: a chemical time bomb? Springer Verlag, Berlin

Debieux CM, Dridge EJ, Mueller CM, Splatt P, Paszkiewicz K, Knight I, Florance H, Love J, Titball RW, Lewis RJ (2011) A bacterial process for selenium nanosphere assembly. Proc Natl Acad Sci USA 108:13480–13485

Dhankher OP, Rosen BP, McKinney EC, Meagher RB (2006) Hyperaccumulation of arsenic in the shoots of Arabidopsis silenced for arsenate reductase (ACR2). Proc Natl Acad Sci USA 103:5413–5418

Dhillon K, Dhillon S, Dogra R (2010) Selenium accumulation by forage and grain crops and volatilization from seleniferous soils amended with different organic materials. Chemosphere 78:548–556

Dostalek P, Patzak M, Matejka P (2004) Influence of specific growth limitation on biosorption of heavy metals by *Saccharomyces cerevisiae*. Int Biodeterior Biodegradation 54:203–207

Douglas S, Beveridge TJ (1998) Mineral formation by bacteria in natural microbial communities. FEMS Microbiol Ecol 26:79–88

Drahota P, Rohovec J, Filippi M, Mihaljevic M, Rychlovský P, Cervený V, Pertold Z (2009) Mineralogical and geochemical controls of arsenic speciation and mobility under different redox conditions in soil, sediment and water at the Mokrsko-West gold deposit, Czech Republic. Sci Total Environ 407:3372–3384

Drexel RT, Haitzer M, Ryan JN, Aiken GR, Nagy KL (2002) Mercury (II) sorption to two Florida Everglades peats: evidence for strong and weak binding and competition by dissolved organic matter released from the peat. Environ Sci Technol 36:4058–4064

Dube A, Zbytniewski R, Kowalkowski T, Cukrowska E, Buszewski B (2001) Adsorption and migration of heavy metals in soil. Pol J Environ Stud 10:1–10

Duester L, Vink JM, Hirner AV (2008) Methylantimony and -arsenic species in sediment pore water tested with the sediment or fauna incubation experiment. Environ Sci Technol 42:5866–5871

Dungan RS, Frankenberger Jr. WT (2000) Factors affecting the volatilization of dimethylselenide by *Enterobacter cloacae* SLD1a-1 Soil Biol Biochem 32:1353–1358

Dursun AY (2006) A comparative study on determination of the equilibrium, kinetic and thermo-dynamic parameters of biosorption of copper (II) and lead (II) ions onto pretreated *Aspergillus niger*. Biochem Eng J 28:187–195

Eary LE, Rai D (1991) Chromate reduction by subsurface soils under acidic conditions. Soil Sci Soc Am J 55:676

Edvantoro BB, Naidu R, Megharaj M, Merrington G, Singleton I (2004) Microbial formation of volatile arsenic in cattle dip site soils contaminated with arsenic and DDT. Appl Soil Ecol 25:207–217

Ehrlich HL (1996) Geomicrobiology, 3rd edn. Dekker, New York

Elbaz-Poulichet F, Dupuy C, Cruzado A, Velasquez Z, Achterberg EP, Braungardt CB (2000) Influence of sorption processes by iron oxides and algae fixation on arsenic and phosphate cycle in an acidic estuary (Tinto river, Spain). Water Res 34:3222–3230

Emett MT, Khoe GH (2001) Photochemical oxidation of arsenic by oxygen and iron in acidic solutions. Water Res 35:649–656

Fergusson JF, Gavis J (1972) A review of the arsenic cycle in natural waters. Water Res 6:1259–1274

Fernández-Martínez A, Charlet L (2009) Selenium environmental cycling and bioavailability: a structural chemist point of view. Rev Environ Sci Biotechnol 8:81–110

Fitz WJ, Wenzel WW (2002) Arsenic transformations in the soil–rhizosphere–plant system: fundamentals and potential application to phytoremediation. J Biotechnol 99:259–278

Fleming EJ, Mack EE, Green PG, Nelson DC (2006) Mercury methylation from unexpected sources: molybdate-inhibited freshwater sediments and an iron-reducing bacterium. Appl Environ Microbiol 72:457–464

Franco DV, Da Silva LM, Jardim WF (2009) Chemical reduction of hexavalent chromium present in contaminated soil using a packed bed column reactor. CLEAN 37:858–865

Frankenberger WT Jr, Arshad M (2001) Bioremediation of selenium contaminated sediments and water. Biofactors 14:241–254

Frankenberger W, Arshad M (2002) Volatilization of arsenic. In: Frankenberger W (ed) Environmental chemistry of arsenic. Marcel Dekker, New York, pp 363–380

Frankenberger WT Jr, Karlson U (1994a) Soil management factors affecting volatilization of selenium from dewatered sediments. Geomicrobiol J 12:265–278

Frankenberger WT Jr, Karlson U (1994b) Microbial volatilization of selenium from soils and sediments. In: Frankenberger WT Jr, Benson S (eds) Selenium in the environment. Marcel Dekker, New York, pp 369–387

Frankenberger WT, Arshad M, Siddique T, Han SK, Okeke BC, Zhang Y (2005) Bacterial diversity in selenium reduction of agricultural drainage water amended with rice straw. J Environ Qual 34:217–226

Frankenberger WT, Losi ME (1995) Application of bioremediation in the cleanup of heavy elements and metalloids. In: Skipper HD, Turco RF (eds) Bioremediation: science and applications, Soil science special publication No. 43. Soil Science Society of America Inc, Madison, WI, pp 173–210

Fulladosa E, Murat JC, Martinez M, Villaescusal I (2004) Effect of pH on arsenate and arsenite toxicity to luminescent bacteria (*Vibrio fischeri*). Arch Environ Contam Toxicol 46:176–182

Gadd G (1993) Microbial formation and transformation of organometallic and organometalloid compounds. FEMS Microbiol Rev 11:297–316

Gadd GM (1990) Heavy metal accumulation by bacteria and other microorganisms. Cell Mol Life Sci 46:834–840

Gadd GM (2008) Fungi and their role in the biosphere. In: Jorgensen SE, Fath B (eds) Encyclopedia of ecology. Elsevier, Amsterdam, pp 1709–1717

Gadd GM (2010) Metals, minerals and microbes: geomicrobiology and bioremediation. Microbiology 156:609–643

Gao S, Tanji KK (1995) Model for biomethylation and volatilization of selenium from agricultural evaporation ponds. J Environ Qual 24:191–197

Geng B, Jin Z, Li T, Qi X (2009) Kinetics of hexavalent chromium removal from water by chitosan-Fe0 nanoparticles. Chemosphere 75:825–830

Geoffrey M, Gadd G (2007) Geomycology: biogeochemical transformations of rocks, minerals, metals and radionuclides by fungi, bioweathering and bioremediation. Mycol Res 111:3–49

Gilmour CC, Elias DA, Kucken AM, Brown SD, Palumbo AV, Schadt CW, Wall JD (2011) Sulfate-reducing bacterium *Desulfovibrio desulfuricans* ND132 as a model for understanding bacterial mercury methylation. Appl Environ Microbiol 77:3938–3951

Gilmour CC, Henry EA (1991) Mercury methylation in aquatic systems affected by acid deposition. Environ Poll 71:131–169

Goh KH, Lim TT (2005) Arsenic fractionation in a fine soil fraction and influence of various anions on its mobility in the subsurface environment. Appl Geochem 20:229–239

Gong C, Donahoe RJ (1997) An experimental study of heavy metal attenuation and mobility in sandy loam soils. Appl Geochem 12:243–254

Graham AM, Aiken GR, Gilmour CC (2012) Dissolved organic matter enhances microbial mercury methylation under sulfidic conditions. Environ Sci Technol 46:2715–2723

Green-Ruiz C (2006) Mercury (II) removal from aqueous solutions by nonviable *Bacillus* sp. from a tropical estuary. Bioresour Technol 97:1907–1911

Guo L, Frankenberger WT Jr, Jury WA (1999) Evaluation of simultaneous reduction and transport of selenium in saturated soil columns. Water Resour Res 35:663–669

Hammerschmidt CR, Fitzgerald WF (2006) Photodecomposition of methylmercury in an arctic Alaskan lake. Environ Sci Technol 40:1212–1216

Hansel CM, Fendorf S, Jardine PM, Francis CA (2008) Changes in bacterial and archaeal community structure and functional diversity along a geochemically variable soil profile. Appl Environ Microbiol 74:1620–1633

Hansen HK, Ribeiro A, Mateus E (2006) Biosorption of arsenic (V) with *Lessonia nigrescens*. Min Eng 19:486–490

Hartley W, Dickinson NM, Riby P, Lepp NW (2009) Arsenic mobility in brownfield soils amended with green waste compost or biochar and planted with *Miscanthus*. Environ Pollut 157:2654–2662

Hasan S, Ranjan D, Talat M (2010) Agro-industrial waste 'wheat bran' for the biosorptive remediation of selenium through continuous up-flow fixed-bed column. J Hazard Mater 181:1134–1142

Haswell SJ, O'Neill P, Bancroft KC (1985) Arsenic speciation in soil-pore waters from mineralized and unmineralized areas of south-west England. Talanta 32:69–72

Haygarth PM, Fowler D, Sturup S, Davison BM, Tones KC (1994) Determination of gaseous and particulate selenium over a rural grassland in the UK. Atmos Environ 28:3655–3663

He Z, Gao F, Sha T, Hu Y, He C (2009) Isolation and characterization of a Cr(VI)-reduction *Ochrobactrum* sp. strain CSCr-3 from chromium landfill. J Hazard Mater 163:869–873

Hechun P, Guangshen L, Zhiyun Y, Yetang H (1996) Acceleration of selenate reduction by alternative drying and wetting of soils. Chin J Geochem 15:278–284

Heeraman D, Claassen V, Zasoski R (2001) Interaction of lime, organic matter and fertilizer on growth and uptake of arsenic and mercury by Zorro fescue (*Vulpia myuros* L.). Plant Soil 234:215–231

Herbel MJ, Blum JS, Oremland RS, Borglin SE (2003) Reduction of elemental selenium to selenide: experiments with anoxic sediments and bacteria that respire Se-oxyanions. Geomicrobiol J 20:587–602

Heyes A, Mason RP, Kim EH, Sunderland E (2006) Mercury methylation in estuaries: insights from using measuring rates using stable mercury isotopes. Mar Chem 102:134–147

Hider RC, Kong X (2010) Chemistry and biology of siderophores. Nat Prod Rep 27:637–657

Higgins TE, Halloran A, Dobbins M, Pittignano A (1998) In situ reduction of hexavalent chromium in alkaline soils enriched with chromite ore processing residue. J Air Waste Manage Assoc 48:1100–1106

Hinsinger P, Bengough G, Vetterlein D, Young IM (2009) Rhizosphere: biophysics, biochemistry and ecological relevance. Plant Soil 321:117–152

Horton RN, Apel WA, Thompson VS, Sheridan PP (2006) Low temperature reduction of hexavalent chromium by a microbial enrichment consortium and a novel strain of *Arthrobacter aurescens*. BMC Microbiol 6:5. doi:10.1186/1471-2180-6-5

Hsu L, Wang S, Lin Y, Wang M, Chiang P, Liu J, Kuan W, Chen C, Tzou Y (2010) Cr(VI) removal on fungal biomass of *Neurospora crassa*: the importance of dissolved organic carbons derived from the biomass to Cr (VI) reduction. Environ Sci Technol 44:6202–6208

Hsu NH, Wang SL, Lin YC, Sheng GD, Lee JF (2009) Reduction of Cr(VI) by crop-residue-derived black carbon. Environ Sci Technol 43:8801–8806

Huang JH, Voegelin A, Pombo SA, Lazzaro A, Zeyer J, Kretzschmar R (2011) Influence of arsenate adsorption to ferrihydrite, goethite, and boehmite on the kinetics of arsenate reduction by *Shewanella putrefaciens* strain CN-32. Environ Sci Technol 44:6202–6208

Ibrahim ASS, El-Tayeb MA, Elbadawi YB, Al-Salamah AA (2011) Isolation and characterization of novel potent Cr(VI) reducing alkaliphilic *Amphibacillus* sp. KSUCr3 from hypersaline soda lakes. Electron J Biotechnol 4:1–14

Ikram M, Faisal M (2010) Comparative assessment of selenite (SeIV) detoxification to elemental selenium (Se^0) by *Bacillus* sp. Biotechnol Lett 32:1255–1259

Iyer A, Mody K, Jha B (2005) Biosorption of heavy metals by a marine bacterium. Mar Pollut Bull 50:340–343

Jackson TA (1989) The influence of clay minerals, oxides, and humic matter on the methylation and demethylation of mercury by micro organisms in freshwater sediments. Appl Organomet Chem 3:1–30

James BR (2001) Remediation-by-reduction strategies for chromate-contaminated soils. Environ Geochem Health 23:175–179

James BR, Bartlett RJ (1983) Behavior of chromium in soils: VII. Adsorption and reduction of hexavalent forms. J Environ Qual 12:177–181

Jardine P, Fendorf S, Mayes M, Larsen I, Brooks S, Bailey W (1999) Fate and transport of hexavalent chromium in undisturbed heterogeneous soil. Environ Sci Technol 33:2939–2944

Jones C, Anderson H, McDermott K, Inskeep T (2000) Rates of microbially mediated arsenate reduction and solubilization. Soil Sci Soc Am J 64:600

Karlson U, Frankenberger WT Jr, Spencer WF (1994) Physico-chemical properties of dimethyl selenide. J Chem Eng Data 39:608–610

Kelly C, Rudd JWM, Holoka M (2003) Effect of pH on mercury uptake by an aquatic bacterium: implications for Hg cycling. Environ Sci Technol 37:2941–2946

Kim JGD, Chusuei JB, Deng CC (2002) Oxidation of chromium(III) to (VI) by manganese oxides. Soil Sci Soc Am J 66:306–315

Kim MJ (2010) Effects of pH, adsorbate/adsorbent ratio, temperature and ionic strength on the adsorption of arsenate onto soil. Geochem Explor Env A 10:407–412

Knauer K, Behra R, Hemond H (1999) Toxicity of inorganic and methylated arsenic to algal communities from lakes along an arsenic contamination gradient. Aquat Toxicol 46:221–230

Kocman D, Horvat M (2010) A laboratory based experimental study of mercury emission from contaminated soils in the River Idrijca catchment. Atmos Chem Phys Discuss 10:1417–1426

Kodukula PS, Patterson JW, Surampalli RY (1994) Sorption and precipitation of metals in activated-sludge. Biotechnol Bioeng 43:874–880

Kosolapov D, Kuschk P, Vainshtein M, Vatsourina A, Wiessner A, Kästner M, Müller R (2004) Microbial processes of heavy metal removal from carbon-deficient effluents in constructed wetlands. Eng Life Sci 4:403–411

Krämer U, Talke IN, Hanikenne M (2007) Transition metal transport. FEBS Lett 581:2263–2272

Kumagai Y, Sumi D (2007) Arsenic: signal transduction, transcription factor, and biotransformation involved in cellular response and toxicity. Annu Rev Pharmacol Toxicol 47:243–262

Lambertsson L, Nilsson M (2006) Organic material: the primary control on mercury methylation and ambient methyl mercury concentrations in estuarine sediments. Environ Sci Technol 40:1822–1829

Landrot G, Ginder-Vogel M, Sparks DL (2009) Kinetics of chromium (III) oxidation by manganese (IV) oxides using quick scanning X-ray absorption fine structure spectroscopy (Q-XAFS). Environ Sci Technol 44:143–149

Lawson S, Macy JM (1995) Bioremediation of selenite in oil refinery waste-water. Appl Microbiol Biotechnol 43:762–765

Ledin M, Krantz-Rulcker C, Allard B (1999) Microorganisms as metal sorbents: comparison with other soil constituents in multi-compartment systems. Soil Biol Biochem 31:1639–1648

Lee DY, Shih YN, Zheng HC, Chen CP, Juang KW, Lee JF, Tsui L (2006) Using the selective ion exchange resin extraction and XANES methods to evaluate the effect of compost amendments on soil chromium(VI) phytotoxicity. Plant Soil 281:87–96

Lehr CR (2003) Microbial methylation and volatilization of arsenic. PhD thesis, Department of chemistry, The University of British Columbia, Canada

Leita L, Margon A, Sinicco T, Mondini C (2011) Glucose promotes the reduction of hexavalent chromium in soil. Geoderma 164:122–127

Lens P, Van Hullebusch E, Astratinei V (2006) Bioconversion of selenate in methanogenic anaerobic granular sludge. J Environ Qual 35:1873–1883

Li HF, McGrath SP, Zhao FJ (2008) Selenium uptake, translocation and speciation in wheat supplied with selenate or selenite. New Phytol 178:92–102

Li RY, Ago Y, Liu WJ, Mitani N, Feldmann J, McGrath SP, Ma JF, Zhao FJ (2009) The rice aquaporin Lsi1 mediates uptake of methylated arsenic species. Plant Physiol 150:2071–2080

Liebert CA, Wireman J, Smith T, Summers AO (1997) Phylogeny of mercury resistance (mer) operons of gram-negative bacteria isolated from the fecal flora of primates. Appl Environ Microbiol 63:1066–1076

Lortie L, Gould W, Rajan S, McCready R, Cheng KJ (1992) Reduction of selenate and selenite to elemental selenium by a *Pseudomonas stutzeri* isolate. Appl Environ Microbiol 58:4042–4044

Loseto LL, Siciliano SD, Lean DRS (2004) Methylmercury production in high Arctic wetlands. Environ Toxicol Chem 23:17–23

Losi M, Amrhein C, Frankenberger W Jr (1994) Factors affecting chemical and biological reduction of hexavalent chromium in soil. Environ Toxicol Chem 13:1727–1735

Losi ME, Frankenberger WT Jr (1997a) Reduction of selenium oxyanions by *Enterobacter cloacae* strain SLD1a 1: reduction of selenate to selenite. Environ Toxicol Chem 16:1851–1858

Losi ME, Frankenberger WT (1997b) Bioremediation of selenium in soil and water. Soil Sci 162:692–702

Loukidou MX, Matis KA, Zouboulis AI, Liakopoulou-Kyriakidou M (2003) Removal of As (V) from wastewaters by chemically modified fungal biomass. Water Res 37:4544–4552

Lovley DR (1995) Bioremediation of organic and metal contaminants with dissimilatory metal reduction. J Ind Microbiol 14:85–93

Mabrouk MEM (2008) Statistical optimization of medium components for chromate reduction by halophilic *Streptomyces* sp. MS-2. Afr J Microbiol Res 2:103–109

Maher W, Butler E (1988) Arsenic in the marine environment. Appl Organomet Chem 2:191–214

Mahimairaja S, Bolan NS, Adriano D, Robinson B (2005) Arsenic contamination and its risk management in complex environmental settings. Adv Agron 86:1–82

Maiers D, Wichlacz P, Thompson D, Bruhn D (1988) Selenate reduction by bacteria from a selenium-rich environment. Appl Environ Microbiol 54:2591–2593

Manning BA, Fendorf SE, Bostick B, Suarez DL (2002) Arsenic(III) oxidation and arsenic (V) adsorption reactions on synthetic birnessite. Environ Sci Technol 36:976–981

Marechal JC, Ahmed S, Engerrand C, Galeazzi L, Touchard F (2006) Threatened groundwater resources in rural India: an example of monitoring. Asian J Water Environ Pollut 3:15–21

Marinari S, Masciandaro G, Ceccanti B, Grego S (2000) Influence of organic and mineral fertilisers on soil biological and physical properties. Bioresour Technol 72:9–17

Marschner H (1995) Mineral nutrition of higher plants, 2nd edn. Academic, London

Marvin-Dipasquale M, Agee J, McGowan C, Oremland RS, Thomas M, Krabbenhoft D, Gilmour CC (2000) Methyl-mercury degradation pathways: a comparison among three mercury-impacted ecosystems. Environ Sci Technol 34:4908–4916

Mason RP, Rolfhus KR, Fitzgerald WF (1995) Methylated and elemental mercury cycling in the surface and deep waters of the North Atlantic. Water Air Soil Pollut 80:665–677

Masscheleyn PH, Delaune RD, Patrick WH Jr (1990) Transformations of selenium as affected by sediment oxidation-reduction potential and pH. Environ Sci Technol 24:91–96

Mehrotra AS, Sedlak DL (2005) Decrease in net mercury methylation rates following iron amendment to anoxic wetland sediment slurries. Environ Sci Technol 39:2564–2570

Mejáre M, Bülow L (2001) Metal-binding proteins and peptides in bioremediation and phytoremediation of heavy metals. Trends Biotechnol 19:67–73

Melo J, D'Souza S (2004) Removal of chromium by mucilaginous seeds of *Ocimum basilicum*. Bioresour Technol 92:151–155

Michalke K, Wickenheiser E, Mehring M, Hirner A, Hensel R (2000) Production of volatile derivatives of metal(loid)s by microflora involved in anaerobic digestion of sewage sludge. Appl Environ Microbiol 66:2791–2796

Miskimmin BM, Rudd JWM, Kelly CA (1992) Influences of DOC, pH, and microbial respiration rates of mercury methylation and demethylation in lake water. Can J Fish Aquat Sci 49:17–22

Miyata N, Tani Y, Sakata M, Iwahori K (2007) Microbial manganese oxide formation and interaction with toxic metal ions. J Biosci Bioeng 104:1–8

Mohanty K, Jha M, Meikap B, Biswas M (2006) Biosorption of Cr (VI) from aqueous solutions by *Eichhornia crassipes*. Chem Eng J 117:71–77

Moreno FN, Anderson CWN, Stewart RB, Robinson BH, Ghomshei M, Meech JA (2005a) Induced plant uptake and transport of mercury in the presence of sulphur containing ligands and humic acid. New Phytol 166:445–454

Moreno FN, Anderson CWN, Stewart RB, Robinson BH, Nomura R, Ghomshei M, Meech JA (2005b) Effect of thioligands on plant-Hg accumulation and volatilisation from mercury-contaminated mine tailings. Plant Soil 275:233–246

Mosher BW, Duce RA (1987) Global atmospheric selenium budget. J Geophys Res 92:13289–13298

Munthe J, Xiao Z, Lindqvist O (1991) The aqueous reduction of divalent mercury by sulfite. Water Air Soil Pollut 56:621–630

Murphy V, Hughes H, McLoughlin P (2008) Comparative study of chromium biosorption by red, green and brown seaweed biomass. Chemosphere 70:1128–1134

Murugesan G, Sathishkumar M, Swaminathan K (2006) Arsenic removal from groundwater by pretreated waste tea fungal biomass. Bioresour Technol 97:483–487

Musante A (2008) The role of mercury speciation in its methylation by methylcobalamin (vitamin-B12). Bachelor thesis, Wheaton College, Norton, MA

Myneni S, Tokunaga TK, Brown GE Jr (1997) Abiotic selenium redox transformations in the presence of Fe(II, III) oxides. Science 278:1106–1109

Nakayasu K, Fukushima M, Sasaki K, Tanaka S, Nakamura H (1999) Comparative studies of the reduction behavior of chromium(VI) by humic substances and their precursors. Environ Toxicol Chem 18:1085–1090

Navratilova J, Raber G, Fisher SJ, Francesconi KA (2011) Arsenic cycling in marine systems: degradation of arsenosugars to arsenate in decomposing algae, and preliminary evidence for the formation of recalcitrant arsenic. Environ Chem 8:44–51

Ndung'u K, Friedrich S, Gonzalez AR, Flegal AR (2010) Chromium oxidation by manganese (hydr) oxides in a California aquifer. Appl Geochem 25:377–381

Newman DK, Beveridge TJ, Morel FMM (1997) Precipitation of As_2S_3 by *Desulfotomaculum auripigmentum*. Appl Environ Microbiol 63:2022–2028

Nies DH (1999) Microbial heavy-metal resistance. Appl Microbiol Biotechnol 51:730–750

Niu CH, Volesky B, Cleiman D (2007) Biosorption of arsenic(V) with acid-washed crab shells. Water Res 41:2473–2478

Oiffer L, Siciliano SD (2009) Methyl mercury production and loss in Arctic soil. Sci Total Environ 407:1691–1700

Oliver DS, Brockman FJ, Bowman RS, Kieft TL (2003) Microbial reduction of hexavalent chromium under vadose zone conditions. J Environ Qual 32:317–324

Opperman DJ, Piater LA, Van Heerden E (2008) A novel chromate reductase from *Thermus scotoductus* SA-01 related to old yellow enzyme. J Bacteriol 190:3076–3082

Oremland RS, Herbel MJ, Blum JS, Langley S, Beveridge TJ, Ajayan PM, Sutto T, Ellis AV, Curran S (2004) Structural and spectral features of selenium nanospheres produced by Se-respiring bacteria. Appl Environ Microbiol 70:52–60

Oremland RS, Hollibaugh JT, Maest AS, Presser TS, Miller LG, Culbertson CW (1989) Selenate reduction to elemental selenium by anaerobic bacteria in sediments and culture: biogeochemical significance of a novel, sulfate-independent respiration. Appl Environ Microbiol 55:2333–2343

Öztürk A, Artan T, Ayar A (2004) Biosorption of nickel(II) and copper(II) ions from aqueous solution by *Streptomyces coelicolor* A3(2). Colloids Surf B Biointerfaces 34:105–111

Pacyna E, Pacyna J, Pirrone N (2001) European emissions of atmospheric mercury from anthropogenic sources in 1995. Atmos Environ 35:2987–2996

Pal S, Vimala Y (2011) Bioremediation of chromium from fortified solutions by *Phanerochaete chrysosporium* (MTCC 787). J Bioremed Biodegradation 2:127

Park D, Yun YS, Park JM (2004) Reduction of hexavalent chromium with the brown seaweed *Ecklonia* biomass. Environ Sci Technol 38:4860–4864

Park JH, Lamb D, Paneerselvam P, Choppala G, Bolan N, Chung JW (2011a) Role of organic amendments on enhanced bioremediation of heavy metal (loid) contaminated soils. J Hazard Mater 185:549–574

Park JH, Bolan NS, Megharaj M, Naidu R (2011b) Isolation of phosphate solubilizing bacteria and their potential for lead immobilization in soil. J Hazard Mater 185:829–836

Park JH, Bolan NS, Megharaj M, Naidu R, Chung JW (2011c) Bacterial-assisted immobilization of lead in soils: implications for remediation. Pedologist 54:162–174

Parvathi K, Nagendran R (2008) Functional groups on waste beer yeast involved in chromium biosorption from electroplating effluent. World J Microbiol Biotechnol 24:2865–2870

Paul J, Beauchamp E (1989) Effect of carbon constituents in manure on denitrification in soil. Can J Soil Sci 69:49–61

Pécheyran C, Quetel CR, Lecuyer FMM, Donard OFX (1998) Simultaneous determination of volatile metal (Pb, Hg, Sn, In, Ga) and nonmetal species (Se, P, As) in different atmospheres by cryofocusing and detection by ICPMS. Anal Chem 70:2639–2645

Pédrot M, Dia A, Davranche M, Bouhnik-Le Coz M, Henin O, Gruau G (2008) Insights into colloid-mediated trace element release at the soil/water interface. J Colloid Interface Sci 325:187–197

Peitzsch M, Kremer D, Kersten M (2010) Microfungal alkylation and volatilization of selenium adsorbed by goethite. Environ Sci Technol 44:129–135

Perpetuo EA, Souza CB, Nascimento CAO (2011) Engineering bacteria for bioremediation. In: Carpi A (ed) Progress in molecular and environmental bioengineering—from analysis and modeling to technology applications. InTech Publishers, Rijeka, pp 605–632

Prasad KS, Srivastava P, Subramanian V, Paul J (2011) Biosorption of As (III) ion on *Rhodococcus* sp. WB-12: biomass characterization and kinetic studies. Separ Sci Technol 46:2517–2525

Qin J, Lehr CR, Yuan C, Le XC, McDermott TR, Rosen BP (2009) Biotransformation of arsenic by a Yellowstone thermoacidophilic eukaryotic alga. Proc Natl Acad Sci USA 106: 5213–5217

Qureshi S, Richards BK, McBride MB, Baveye P, Steenhuis TS (2003) Temperature and microbial activity effects on trace element leaching from metalliferous peats. J Environ Qual 32:2067–2075

Ramial P, John WMR, Furutam A, Xun L (1985) The effect of pH on methyl mercury production and decomposition in lake sediments. Can J Fish Aquat Sci 42:685–692

Ranjard L, Prigent-Combaret C, Nazaret S, Cournoyer B (2002) Methylation of inorganic and organic selenium by the bacterial thiopurine methyltransferase. J Bacteriol 184:3146–3149

Ravichandran M (2004) Interactions between mercury and dissolved organic matter—a review. Chemosphere 55:319–331

Reategui M, Maldonado H, Ly M, Guibal E (2010) Mercury(II) biosorption using *Lessonia* sp Kelp. Appl Biochem Biotechnol 162:805–822

Rech S, Macy J (1992) The terminal reductases for selenate and nitrate respiration in *Thauera selenatis* are two distinct enzymes. J Bacteriol 174:7316–7320

Regnell O, Tunlid A (1991) Laboratory study of chemical speciation of mercury in lake sediment and water under aerobic and anaerobic conditions. Appl Environ Microbiol 57:789–795

Rendina A, Barros M, de Iorio A (2006) Phytoavailability and solid-phase distribution of chromium in a soil amended with organic matter. Bull Environ Contam Toxicol 76:1031–1037

Renshaw JC, Robson GD, Trinci APJ, Wiebe MG, Livens FR, Collison D, Taylor RJ (2002) Fungal siderophores: structures, functions and applications. Mycol Res 106:1123–1142

Robinson B, Bolan NS, Mahimairaja S, Clothier B (2006) Solubility, mobility and bioaccumulation of trace elements: abiotic processes in the rhizosphere. In: Prasad M, Sajwan K, Naidu R (eds) Trace elements in the environment: biogeochemistry, biotechnology and bioremediation. CRC Press, London, pp 97–110

Rocha JC, Junior ÉS, Zara LF, Rosa AH, dos Santos A, Burba P (2000) Reduction of mercury (II) by tropical river humic substances (Rio Negro)—A possible process of the mercury cycle in Brazil. Talanta 53:551–559

Rochette EA, Bostick BC, Li GC, Fendorf S (2000) Kinetics of arsenate reduction by dissolved sulfide. Environ Sci Technol 34:4714–4720

Rock ML, James BR, Helz GR (2001) Hydrogen peroxide effects on chromium oxidation state and solubility in four diverse, chromium-enriched soils. Environ Sci Technol 35:4054–4059

Rodríguez Martín-Doimeadios R, Tessier E, Amouroux D, Guyoneaud R, Duran R, Caumette P, Donard O (2004) Mercury methylation/demethylation and volatilization pathways in estuarine sediment slurries using species-specific enriched stable isotopes. Mar Chem 90:107–123

Roeselers G, van Loosdrecht MCM, Muyzer G (2008) Phototrophic biofilms and their potential applications. J Appl Phycol 20:227–235

Rogers R (1976) Methylation of mercury in agricultural soils. J Environ Qual 5:454–458

Rogers RD, MacFarlane JC (1978) Factors influencing the volatilization of mercury from soil. Environmental Protection Agency, Office of Research and Development, Environmental Monitoring and Support Laborator, Las Vegas

Rosen BP, Silver S (1987) Ion transport in prokaryotes. Academic, San Diego, CA

Ross SM (1994) Retention, transformation and mobility of toxic metals in soils. In: Ross SM (ed) Toxic metals in soil–plant systems. Wiley, New York, pp 63–152

Roy V, Amyot M, Carignan R (2009) Beaver ponds increase methylmercury concentrations in Canadian shield streams along vegetation and pond-age gradients. Environ Sci Technol 43:5605–5611

Rubinos DA, Iglesias L, Díaz-Fierros F, Barral MT (2011) Interacting effect of ph, phosphate and time on the release of arsenic from polluted river sediments (Anllóns River, Spain). Aquat Geochem 17:281–306

Sadiq M (1997) Arsenic chemistry in soils: an overview of thermodynamic predictions and field observations. Water Air Soil Pollut 93:117–136

Sağlam N, Say R, Denizli A, Patır S, Arıca MY (1999) Biosorption of inorganic mercury and alkyl-mercury species on to Phanerochaete chrysosporium mycelium. Process Biochem 34:725–730

Salomons W, Stigliani W (1995) Biogeodynamics of pollutants. Springer, Berlin, p 257

Sand W, Gehrke T, Jozsa PG, Schippers A (2001) (Bio)chemistry of bacterial leaching-direct vs. indirect bioleaching. Hydrometallurgy 59:159–175

Sass H, Ramamoorthy S, Yarwood C, Langner H, Schumann P, Kroppenstedt R, Spring S, Rosenzweig R (2009) Desulfovibrio idahonensis sp. nov., sulfate-reducing bacteria isolated from a metal (loid)-contaminated freshwater sediment. Int J Syst Evol Microbiol 59:2208–2214

Schiewer S, Volesky B (2000) Biosorption processes for heavy metal removal. In: Lovley DR (ed) Environmental microbe-metal interactions. ASM Press, Washington, DC, pp 329–362

Schlüter K (2000) Review: evaporation of mercury from soils. An integration and synthesis of current knowledge. Environ Geol 39:249–271

Schroeder WH, Munthe J (1998) Atmospheric mercury—an overview. Atmos Environ 32:809–822

Schwesig D, Matzner E (2001) Dynamics of mercury and methylmercury in forest floor and runoff of a forested watershed in Central Europe. Biogeochemistry 53:181–200

Sharma S, Bansal A, Dogra R, Dhillon SK, Dhillon KS (2011) Effect of organic amendments on uptake of selenium and biochemical grain composition of wheat and rape grown on seleniferous soils in northwestern India. J Plant Nutr Soil Sci 174:269–275

Shrestha B, Lipe S, Johnson KA, Zhang TQ, Retzlaff W, Lin ZQ (2006) Soil hydraulic manipulation and organic amendment for the enhancement of selenium volatilization in a soil-pickleweed system. Plant Soil 288:189–196

Singh G, Brar M, Malhi S (2007) Decontamination of chromium by farm yard manure application in spinach grown in two texturally different Cr-contaminated soils. J Plant Nutr 30:289–308

Skopp J, Jawson MD, Doran JW (1990) Steady-state aerobic microbial activity as a function of soil water content. Soil Sci Soc Am 54:1619–1625

Smith E, Naidu R, Alston AM (1998) Arsenic in the soil environment: a review. Adv Agron 66:149–195

Smith WA, Apel WA, Petersen JN, Peyton BM (2002) Effect of carbon and energy source on bacterial chromate reduction. Biorem J 6:205–215

Smolders E, Buekers J, Oliver I, McLaughlin MJ (2004) Soil properties affecting toxicity of zinc to soil microbial properties in laboratory-spiked and field contaminated soils. Environ Toxicol Chem 23:2633–2640

Song X, Heyst BV (2005) Volatilization of mercury from soils in response to simulated precipitation. Atmos Environ 39:7494–7505

Srinath T, Verma T, Ramteke P, Garg S (2002) Chromium(VI) biosorption and bioaccumulation by chromate resistant bacteria. Chemosphere 48:427–435

Sukumar M (2010) Reduction of hexavalent chromium by *Rhizopus Oryzae*. Afr J Environ Sci Technol 4:412–418

Sun X, Wang Q, Ma H, Wang Z, Yang S, Zhao C, Xu L (2011) Effects of plant rhizosphere on mercury methylation in sediments. J Soils Sediments 11:1062–1069

Surowitz KG, Titus JA, Pfister RM (1984) Effects of cadmium accumulation on growth and respiration of a cadmium-sensitive strain of *Bacillus subtilis* and a selected cadmium resistant mutant. Arch Microbiol 140:107–112

Suseela K, Sivaparvathi M, Nandy SC (1987) Removal of chromium from tannery effluent using powdered leaves. Leather Sci (Madras) 34:149–156

Svecova L, Spanelova M, Kubal M, Guibal E (2006) Cadmium, lead and mercury biosorption on waste fungal biomass issued from fermentation industry. I. Equilibrium studies. Sep Purif Technol 52:142–153

Tan T, Beydoun D, Amal R (2003) Effects of organic hole scavengers on the photocatalytic reduction of selenium anions. J Photochem Photobiol A 159:273–280

Tandukar M, Huber SJ, Onodera T, Pavlostathis SG (2009) Biological chromium (VI) reduction in the cathode of a microbial fuel cell. Environ Sci Technol 43:8159–8165

Tarze A, Dauplais M, Grigoras I, Lazard M, Ha-Duong NT, Barbier F, Blanquet S, Plateau P (2007) Extracellular production of hydrogen selenide accounts for thiol-assisted toxicity of selenite against *Saccharomyces cerevisiae*. J Biol Chem 282:8759–8767

Templeton AS, Trainor TP, Spormann AM, Brown GE Jr (2003) Selenium speciation and partitioning within *Burkholderia cepacia* biofilms formed on $\alpha\text{-}Al_2O_3$ surfaces. Geochim Cosmochim Acta 67:3547–3557

Thayer JS, Brinckman FE (1982) The biological methylation of metals and metal-loids. Adv Organomet Chem 20:313–356

Thompson-Eagle E, Frankenberger WT Jr, Karlson U (1989) Volatilization of selenium by *Alternaria alternata*. Appl Environ Microbiol 55:1406–1413

Thompson-Eagle ET, Frankenberger WT Jr (1990) Site volatilization of selenium with alternative sources of protein for microbial deselenification at evaporation ponds. J Environ Qual 19:125–129

Thompson-Eagle ET, Franakenberger WT Jr (1992) Bioremediation of soils contaminated with selenium. In: Lal R, Stewart BA (eds) Advances in soil science. Springer, New York, pp 261–310

Tseng JK, Bielefeldt AR (2002) Low-temperature chromium (VI) biotransformation in soil with varying electron acceptors. J Environ Qual 31:1831–1841

Tüzen M, Özdemir M, Demirbaş A (1998) Heavy metal bioaccumulation by cultivated *Agaricus bisporus* from artificially enriched substrates. Z Lebensm Unters Forsch 206:417–419

Tuzen M, Sari A (2010) Biosorption of selenium from aqueous solution by green algae (*Cladophora hutchinsiae*) biomass: equilibrium, thermodynamic and kinetic studies. Chem Eng J 158:200–206

Ucun H, Bayhan YK, Kaya Y, Cakici A, Faruk Algur O (2002) Biosorption of chromium (VI) from aqueous solution by cone biomass of *Pinus sylvestris*. Bioresour Technol 85:155–158

Ullrich SM, Tanton TW, Abdrashitova SA (2001) Mercury in the aquatic environment: a review of factors affecting methylation. Crit Rev Environ Sci Technol 31:241–293

Vainshtein M, Kuschk P, Mattusch J, Vatsourina A, Wiessner A (2003) Model experiments on the microbial removal of chromium from contaminated groundwater. Water Res 37:1401–1405

Vera SM, Werth CJ, Sanford RA (2001) Evaluation of different polymeric organic materials for creating conditions that favor reductive processes in groundwater. Biorem J 5:169–181

Viamajala S, Peyton BM, Apel WA, Petersen JN (2002) Chromate/nitrite interactions in *Shewanella oneidensis* MR 1: evidence for multiple hexavalent chromium [Cr (VI)] reduction mechanisms dependent on physiological growth conditions. Biotechnol Bioeng 78:770–778

Vieira M, Oisiovici R, Gimenes M, Silva M (2008) Biosorption of chromium(VI) using a *Sargassum* sp. packed-bed column. Bioresour Technol 99:3094–3099

Volesky B, Holan Z (1995) Biosorption of heavy metals. Biotechnol Prog 11:235–250

Von Canstein H, Kelly S, Li Y, Wagner-Döbler I (2002) Species diversity improves the efficiency of mercury-reducing biofilms under changing environmental conditions. Appl Environ Microbiol 68:2829–2837

Wang D, Qing C, Guo T, Guo Y (1997) Effects of humic acid on transport and transformation of mercury in soil-plant systems. Water Air Soil Pollut 95:35–43

Wang J, Feng X, Anderson CWN, Xing Y, Shang L (2012) Remediation of mercury contaminated sites—a review. J Hazard Mater 221–222:1–18

Wang S, Mulligan CN (2006) Natural attenuation processes for remediation of arsenic contaminated soils and groundwater. J Hazard Mater 138:459–470

Wang YP, Shi JY, Wang H, Lin Q, Chen XC, Chen YX (2007) The influence of soil heavy metals pollution on soil microbial biomass, enzyme activity, and community composition near a copper smelter. Ecotox Environ Safety 67:75–81

Wardle DA (1992) A comparative assessment of factors which influence microbial biomass carbon and nitrogen levels in soil. Biol Rev 67:321–358

Warwick P, Inam E, Evans N (2005) Arsenic's interaction with humic acid. Environ Chem 2:119–124

Watras CJ, Bloom NS (1992) Mercury and methyl mercury in individual zooplankton: implications for bioaccumulation. Limnol Oceanogr 37:1313–1318

Weber FA, Hofacker A, Voegelin A, Kretzschmar A (2010) Temperature dependence and coupling of iron and arsenic reduction and release during flooding of a contaminated soil. Environ Sci Technol 44:116–122

Whalin L, Kim EH, Mason R (2007) Factors influencing the oxidation, reduction, methylation and demethylation of mercury species in coastal waters. Mar Chem 107:278–294

White C, Wilkinson SC, Gadd GM (1995) The role of microorganisms in biosorption of toxic metals and radionuclides. Int Biodeterior Biodegradation 35:17–40

Wiatrowski HA, Ward PM, Barkay T (2006) Novel reduction of mercury (II) by mercury-sensitive dissimilatory metal reducing bacteria. Environ Sci Technol 40:6690–6696

Wiener JG, Gilmour CC, Krabbenhoft DP (2003) Mercury strategy for the bay-delta ecosystem: a unifying framework for science, adaptive management, and ecological restoration. Report to the California Bay Delta authority, Sacramento, California, USA

Wilkie JA, Hering JG (1998) Rapid oxidation of geothermal arsenic(III) in streamwaters of the eastern Sierra Nevada. Environ Sci Technol 32:657–662

Williams JW, Silver S (1984) Bacterial resistance and detoxification of heavy metals. Enzyme Microb Technol 12:530–537

Wu L (2004) Review of 15 years of research on ecotoxicology and remediation of land contaminated by agricultural drainage sediment rich in selenium. Ecotoxicol Environ Saf 57:257–269

Xu X, McGrath S, Zhao F (2007) Rapid reduction of arsenate in the medium mediated by plant roots. New Phytol 176:590–599

Yadav SK, Juwarkar AA, Kumar GP, Thawale PR, Singh SK, Chakrabarti T (2009) Bioaccumulation and phyto-translocation of arsenic, chromium and zinc by *Jatropha curcas* L.: impact of dairy sludge and biofertilizer. Bioresour Technol 100:4616–4622

Yamamura S, Watanabe M, Kanzaki M, Soda S, Ike M (2008) Removal of arsenic from contaminated soils by microbial reduction of arsenate and quinone. Environ Sci Technol 42:6154–6159

Yang T, Chen ML, Hu XW, Wang ZW, Wang JH, Dasgupta PK (2010) Thiolated eggshell membranes sorb and speciate inorganic selenium. Analyst 136:83–89

Yavuz H, Denizli A, Gungunes H, Safarikova M, Safarik I (2006) Biosorption of mercury on magnetically modified yeast cells. Sep Purif Technol 52:253–260

Yin XX, Chen J, Qin J, Sun GX, Rosen BP, Zhu YG (2011) Biotransformation and volatilization of arsenic by three photosynthetic cyanobacteria. Plant Physiol 156:1631–1638

Yin Y, Impellitteri CA, You SJ, Allen HE (2002) The importance of organic matter distribution and extract soil: solution ratio on the desorption of heavy metals from soils. Sci Total Environ 287:107–119

Yoshinaga M, Cai Y, Rosen BP (2011) Demethylation of methylarsonic acid by a microbial community. Environ Microbiol 13:1205–1215

Yun YS, Park D, Park JM, Volesky B (2001) Biosorption of trivalent chromium on the brown seaweed biomass. Environ Sci Technol 35:4353–4358

Zayed AM, Terry N (1994) Selenium volatilization in roots and shoots: effects of shoot removal and sulfate level. J Plant Physiol 143:8–14

Zazo JA, Paull JS, Jaffe PR (2008) Influence of plants on the reduction of hexavalent chromium in wetland sediments. Environ Pollut 156:29–35

Zeng F, Chen S, Miao Y, Wu F, Zhang G (2008) Changes of organic acid exudation and rhizosphere pH in rice plant under chromium stress. Environ Pollut 155:284–289

Zeroual Y, Moutaouakkil A, Zohra Dzairi F, Talbi M, Ung Chung P, Lee K, Blaghen M (2003) Biosorption of mercury from aqueous solution by *Ulva lactuca* biomass. Bioresour Technol 90:349–351

Zhang H, Lindberg SE, Marsik FJ, Keeler GJ (2001) Mercury air/surface exchange kinetics of background soils of the Tahquamenon river watershed in the Michigan Upper Peninsula. Water Air Soil Pollut 126:151–169

Zhang J, Bishop PL (2002) Stabilization/solidification (S/S) of mercury-containing wastes using reactivated carbon and Portland cement. J Hazard Mater 92:199–212

Zhang T, Hsu-Kim H (2010) Photolytic degradation of methylmercury enhanced by binding to natural organic ligands. Nat Geosci 3:473–476

Zhang Y, Frankenberger WT (2003) Factors affecting removal of selenate in agricultural drainage water utilizing rice straw. Sci Total Environ 305:207–216

Zhang YQ, Frankenberger WT (1999) Effects of soil moisture, depth, and organic amendments on selenium volatilization. J Environ Qual 28:1321–1326

Zouboulis A, Loukidou M, Matis K (2004) Biosorption of toxic metals from aqueous solutions by bacteria strains isolated from metal-polluted soils. Process Biochem 39:909–916

Biological Effects of Boron

Mustafa Kabu and Murat Sirri Akosman

Contents

1 Introduction

Boron, which bears the symbol B in the periodical table, is a semiconductive element with properties between that of a metal and a nonmetal (Kılıç et al. 2009). This micro-mineral is ingested with foods on a daily basis, and the amount taken in varies with the levels that occur in the consumed food and drink (Sabuncuoglu et al. 2006).

M. Kabu (✉)
Department of Internal Medicine, Faculty of Veterinary Medicine,
Afyon Kocatepe University, ANS Campus, 03200 Afyonkarahisar, Turkey
e-mail: mkabu@aku.edu.tr

M.S. Akosman
Department of Anatomy, Faculty of Veterinary Medicine, Afyon Kocatepe University,
ANS Campus, 03200 Afyonkarahisar, Turkey

D.M. Whitacre (ed.), *Reviews of Environmental Contamination and Toxicology*,
Reviews of Environmental Contamination and Toxicology 225,
DOI 10.1007/978-1-4614-6470-9_2, © Springer Science+Business Media New York 2013

This element is a chemically dynamic trace element that forms approximately 230 compounds, generally with other elements (World Health Organization 1998; Kılıç et al. 2009).

Boron exists at high concentrations in sedimentary rocks, soils, coal, and seawater (Samman et al. 1998). It is estimated that the global average concentration of boron in seawater is approximately 4.6 mg/L (Samman et al. 1998). Boron is released into the atmosphere from commercial uses, forest fires, coal combustion, and volcanoes. It reaches the ocean as a result of rock weathering, which constitutes another atmospheric source (Howe 1998). Sixty-five to eighty-five percent of boron in the atmosphere is derived from the world's oceans (Argust 1998).

Boron is a component of several manufactured goods, such as glass, detergents, ceramics, and fertilizers, and may reach the environment as a result of being released from these materials or during their production (Argust 1998). Seven to eighteen percent of environmental B derives from several major "anthropogenic" sources, viz., fertilizers, wastewater treatment plant releases, and fly ash waste released by coal-fired power plants (Howe 1998).

Boron is primarily a natural product and generally occurs in the environment as borates (Howe 1998). Borates are boron–oxygen compounds that result from the binding of boron with oxygen. When administered to animals, inorganic borates are biotransformed into boric acids and are absorbed from mucosal surfaces. More than 90% of the borate administered to humans or animals is excreted as boric acid. In both in vitro and in vivo systems, boric acid shows an affinity for *cis*-hydroxyl groups, and such affinity may account for the mechanism by which boric acid produces some of its biological effects (World Health Organization 1998; Bolaños et al. 2004).

Recent studies on the biological significance of boron to various metabolic, nutritional, hormonal, and physiological processes indicated that B may (Blevins and Lukaszewski 1998) or may not (Loomis and Durst 1992) be essential to plants, but B is essential for humans and animals (Nielsen 1997; Basoglu et al. 2000, 2002; Kabu and Civelek 2012; Hunt 2012). It is accepted that boron performs functions in mineral metabolism, in immune response, and in the endocrine system. Furthermore, boron is metabolically important for bone growth and health (Nielsen 1997; Basoglu et al. 2000, 2002; Kabu and Civelek 2012; Hunt 2012). Unfortunately, the detailed mechanism by which boron functions in animals has not yet been fully described.

Our purpose in this chapter is to summarize the current status of knowledge on how boron functions metabolically in living organisms, and produces effects on living organisms.

2 The Importance of Boron for Living Organisms

2.1 Animals

It is now known that boron is a necessary dietary component for humans and animals (Hunt 1994; Nielsen 1997; Kabu and Civelek 2012). Boron meets most criteria as an essential nutrient (Hunt 1998). It has a low atomic weight and binds to organic compounds in ways that influence biological function (Hunt 1998). At the levels boron normally exists in organisms, it is generally nontoxic, and animals tend to have a natural ability to maintain homeostatic control of boron levels in their bodies (Hunt 1998). Notwithstanding, even rather low levels of dietary boron intake in some animal species have been associated with developmental abnormalities (Hunt 1998).

Several studies have been performed to investigate the effects of boron intake on animals. Researchers believe that sodium borate ($Na_2B_4O_7$) protects against developing a fatty liver (Basoglu et al. 2002; Bobe et al. 2004). Because treating the fatty-liver condition in cows is costly and difficult to perform, preventing the disease is a much better approach than having to treat for it after it occurs (Bobe et al. 2004). In one study on cows, significant decreases were observed in serum triglyceride (TG) and very low density lipoprotein (VLDL) levels of animals treated orally with sodium borate (Basoglu et al. 2002). Kabu and Civelek (2012) studied the effects of sodium borate ($Na_2B_4O_7 \cdot 5H_2O$) orally administrated to 12 pregnant cattle at 30 g/day over a 28-day period that included 2-week prepartum and 2-week postpartum exposures. In this study, the effects of sodium borate on selected hormone levels and serum metabolites were investigated in both treated and control animals. Blood samples were obtained weekly. Results were that no differences vs. controls were recorded in blood for concentration of total protein (TP), albumin (ALB), blood urea nitrogen (BUN), alanine aminotransferase (ALT), total bilirubin (TBil), aspartate aminotransferase (AST), and gamma-glutamyltransferase (GGT). Glucose levels were higher during the prepartum period, and the postpartum glucagon and β-hydroxybutyric acid (BHBA) serum levels were higher in the control group (Kabu and Civelek 2012). At the end of sodium borate administration, concentrations of total cholesterol (TChol), triglyceride (TG), high-density lipoprotein (HDL), low-density lipoprotein (LDL), very low density lipoprotein (VLDL), glucose, insulin, and nonesterified fatty acids (NEFA) in blood were decreased (Kabu and Civelek 2012). In summary, administering sodium borate may improve the metabolic situation during the periparturient period (Kabu and Civelek 2012), and the use of sodium borate in cattle during the early lactation period reduces the incidence of fatty liver (Basoglu et al. 2002).

There is evidence that, in some manner that is not yet known, boron balances harmful liver effects by altering oxidative stress parameters and acts to return the liver to its normal level of function (Pawa and Ali 2006). In another study that addressed the effects of B on fatty liver, New Zealand Rabbits were investigated (Basoglu et al. 2010, 2011). The rabbits were administered boron orally at doses of 10, 30, and 50 mg/kg of body weight (Boraks deka hidrat $Na_2B_4O_7 \cdot 10H_2O$) at 96-h intervals (Basoglu et al. 2010). These exposure levels did not affect hematological parameters of the rabbits (Basoglu et al. 2010). Basoglu et al. (2010) suggested that boron has positive effects on hepatic steatosis and visceral fat by reducing oxidative stress and by affecting the lipid profile, although the dose of 50 mg/kg had no stabilizing effects (Basoglu et al. 2010). The author concluded that boron apparently prevents fatty liver by acting on mitochondria. There was evidence in this study that boron affects the Krebs cycle, the glucose–alanine cycle, and methionine metabolism, all of which reduce oxidative stress and positively affect the lipid profile (Basoglu et al. 2011).

Basoglu et al. (2000) fed dogs a daily diet of 4 g/day of borax ($Na_2B_4O_7 \cdot 10H_2O$). That administration level was effective in keeping plasma lipid levels of the dogs low. One week after oral administration of borax, a decrease of glucose, insulin, and apolipoprotein B-100 (Apo-B100) levels was detected in treated dogs vs. controls. A decrease in VLDL and TG levels was also seen after the second week of exposure (Basoglu et al. 2000). These findings supported the conclusion that borax exposure reduced blood lipid levels (Basoglu et al. 2000).

In other studies, exposure of chicks to boron alone increased plasma glucose concentrations, particularly when a vitamin D deficiency existed (Simon and Rosselin 1978; Hunt 1989; Hunt and Herbel 1993). However, in chicks, supplemental dietary boron improved the vitamin D_3 deficiency-induced eleva- tions that existed in plasma glucose concentrations (Hunt et al. 1994). Moreover, in vitamin D-deficient chicks, abnormally elevated plasma concentrations of pyruvate, BHBA, and triglycerides existed that are typical of vitamin D defi- ciency; the addition of dietary boron mitigated these effects (Hunt et al. 1994). In rats, a deficiency of boron produced vitamin D deficiency, which similarly decreased plasma TG concentrations and increased plasma pyruvate concentra- tions (Hunt and Herbel 1991–1992). Boron had no such effect when the diet con- tained sufficient amounts of vitamin D (2,500 IU/kg) (Hunt and Nielsen 1981; Hunt and Herbel 1993; Hunt 1989, 1994). How boron deficiency affects energy substrate metabolism in animal models is unknown, and this is particularly true when there are suboptimal amounts of food intake (Bakken and Hunt 2003). Boron deficiency has also caused hyperinsulinemia in rats that were deprived of vitamin D (Hunt and Herbel 1991–1992). It is claimed that an absence of boron increased the amount of insulin required to maintain plasma glucose concentra- tions, when either vitamin D or magnesium nutriture was perturbed in chicks and rats (Bakken and Hunt 2003).

Hunt et al. (1983) suggested that boron affects vitamin D_3 metabolism or vita- min D_3's effect on growth. DeLuca and Schnoes (1983) manipulated the dietary concentrations of magnesium or calcium to examine the interaction that occurs between dietary vitamin D_3 intake and boron levels. Magnesium deficiency was chosen as a stressor of vitamin D_3 metabolism, because it is a cofactor for the hydroxylation of 25-hydroxycholecalciferol (OH) vitamin D_3 (DeLuca and Schnoes 1983). In another study, boron added to the diet of chickens at a level of 3 mg/kg induced magnesium deficiency (300 mg/kg) and fostered bird growth (Hunt and Nielsen 1981).

The basal model daily diet for chickens contains 10% clover (4.2 mg boron/kg), lab rat daily diets contain 12–13.7 mg boron/kg (Hunt et al. 1988), and human veg- etarians consume daily diets containing at least 2 mg boron/kg (Hunt et al. 1991). Yet, the amount of boron in the basal diets of research animals, in studies carried out before 1950, generally had either insufficient or excessive boron levels (100– 2,200 mg/kg) (Hunt 1998). If boron is added as a supplement to diets of severely potassium-deficient rats that contain levels of 100–1,000 mg boron/kg, a positive effect on survival and maintenance of body fat and elevated liver glycogen resulted (Hunt 1994).

2.2 Humans

The major source of boron entry into the human body is via consumed food. High boron levels exist in fruits, vegetables, pulses, legumes, and nuts, whereas other foods have lower levels. Humans may also acquire boron by consuming certain beverages, or by absorbing B by respiration or through the skin (Nielsen 1997). Boron, once in the body, may be rapidly excreted via urine and does not accumulate in tissues. Boron exists in body tissues and fluids as boric acid (B (OH)$_3$) (Sutherland et al. 1998). In healthy people, total boron blood concentrations are in the range of 15.3–79.5 ng/g wet wt (Clark et al. 1987) and exist as 98.4% boric acid and 1.6% as borate anion (B(OH)$_4$) (Sutherland et al. 1998; Nielsen 1997). Boron content of the various bodily organs varies; values for the heart, liver, lung, kidney, and brain gave levels of 28, 2.31, 0.6, 0.6, 0.06 ppm boron, respectively (Hamilton et al. 1972; Indraprasit et al. 1974; Massie et al. 1990; Nielsen 1997). This suggests that any function carried out by boron differs from organ to organ (Newnham 1991).

Boron can affect the function or composition of the brain, and the skeletal and immune systems (Nielsen 1997). Dietary intake of a daily amount of boron (viz., 3 mg) for more than 7 weeks may prevent osteoporosis in postmenopausal women. After boron was consumed as a supplement, the urinary excretion of the essential elements, calcium and magnesium, decreased by 40% and 33%, respectively (Nielsen et al. 1987). In another study (Nielsen 1990), dietary boron produced effects similar to estrogen supplementation in women who suffer from postmeno-pausal osteoporosis. Boron intake also increased amounts of ionized calcium in the blood serum, decreased serum calcitonin concentrations, and increased serum levels of 1,25-dihydroxycholecalciferol. Dietary boron helps maintain serum calcitonin level, which assists postmenopausal women who suffer from osteoporosis (Nielsen 1990). Penland (1998) performed a study in which inadequate dietary boron intake lowered performance on cognitive and motor tests.

In human males, an orally administered daily dose of 10 mg/day of boron as sodium tetraborate for seven days produced a significant decrease in sex hormone globulin binding (SHGB), high sensitive CRP (hsCRP), and TNF-α levels. Moreover, the mean free testosterone levels in blood plasma increased, and plasma estradiol was significantly decreased (Naghii et al. 2011).

The average amount of boron taken in by humans on a daily basis varies by gender and age. Infants aged 0–6 months, males aged 51–70 years, and lactating females consumed 0.75±0.14 mg/day, 1.34±0.02 mg/day, and 1.39±0.16 mg/day, respectively (Hunt 2012). It was determined that the average daily boron intake in adult males was 1.21 mg/day or 1.52 mg/day (Iyengar et al. 1990; Anderson et al. 1994).

2.3 Plants

Boron is an essential micronutrition for higher plants (Blevins and Lukaszewski 1998). Boron is important in sugar transport, cell wall synthesis and lignification, cell wall structure, carbohydrate metabolism, RNA metabolism, respiration, indole acetic acid metabolism, phenol metabolism, and membrane transport (Blevins and Lukaszewski 1994; Camacho-Cristóbal et al. 2008). The significance of boron to cell wall structure and membrane function is particularly important (Blevins and Lukaszewski 1994; Camacho-CristÓbal et al. 2008).

Fruits, vegetables, and hazelnuts are known to be primary sources of boron (Hunt et al. 1991). Among vegetables, leafy greens have the highest boron levels, especially when they are grown without chemical fertilizers (Newnham 1977). Vegetables, fruits, legumes, and tubers have much higher amounts of boron than do the grasses (e.g., wheat, rice, and corn retain <0.2 mg/kg) (Nielsen 1988; Vanderpool and Johnson 1992). Dried legumes, fruits, avocados, and nuts contain from 1.0 mg to 4.5 mg boron/100 g (Naghii 1999). Fresh fruits, vegetables, honey, and bee pollen contain from 0.1 to 0.6 mg boron/100 g, whereas foods from animal sources have boron at levels between 0.01 and 0.06 mg/100 g (Newnham 1977; Naghii 1999).

The position of the hydroxyl group on the boron atom makes formation of complexes with substrates and other reactants easier (Dugger 1983). Boron may have a significant role in adjusting or regulating certain metabolic functions in plants (Hunt 1994). From studies performed on higher plants (Lovatt and Dugger 1984; Goldbach 1997) and in animals in which human nutrition was evaluated, the position of the hydroxyl group on the boron appeared to be critical (Nielsen et al. 1988; Hunt 1989; Hegsted et al. 1991; Hunt and Herbel 1991–1992, 1993; Bai and Hunt 1996; Eckhert 1998; Fort et al. 1999; Armstrong et al. 2000). Boron may affect metabolic pathways by binding apoplastic proteins to *cis*-hydroxyl groups of cell walls and membranes, and by interfering with manganese-dependent enzymatic reactions. Recently, the formation of borate esters with hydroxyl groups of cell wall carbohydrates and/or glycoproteins has been proposed as a mechanism for cross-linking cell wall polymers (Loomis and Durst 1992). Borate bridging could explain many of the characteristics observed in boron-deficient plants (Blevins and Lukaszewski 1998).

Boron deficiency in plants causes carbohydrate accumulation in chloroplasts, accelerates the activity of the pentose phosphate cycle, and may slow the Krebs cycle (Goldbach 1997; Lovatt and Dugger 1984). The amount of boron available in soils varies with the fertilizers used, and with soil type, temperature, and pH (Newnham 1977). The application of potash fertilizers and superphosphate fertilizers inhibits boron absorption (Newnham 1977). Warm, moist soil renders boron more bioavailable because of increased microfloral activity (Newnham 1977).

2.4   Bacteria and Fungi

Borates produce effects on a variety of bacteria and fungi (Woods 1994; Hunt 2003; Kartal et al. 2004; Rolshausen and Gubler 2005; Baker et al. 2009; Tamay-Cach et al. 2012; Hunt 2012). Hunt (2003) suggests that boron is an essential trace element for at least some organisms in each of the following taxa: Eubacteria, Stramenopila (brown algae and diatoms), Viridiplantae (green algae and familiar green plants), Fungi, and Animalia.

Ahmed and Fujiwara (2010) reported that B is toxic to living cells at levels above a certain threshold. They isolated several B-tolerant bacterial strains from soil samples and studied their possible mechanisms of tolerance to B. They sequenced the gene16S rRNA and performed comparative phylogenetic analysis, which showed that the isolates they studied belonged to the following six genera: *Arthrobacter*, *Rhodococcus*, *Lysinibacillus*, *Algoriphagus*, *Gracilibacillus*, and *Bacillus*. These isolates exhibited tolerance levels to B of 80, 100, 150, 300, 450, and 450 mmol/L, respectively, while maintaining a significantly lower intracellular B concentration than in the medium. Statistical analysis demonstrated a negative correlation between the protoplasmic B concentration and the degree of tolerance to a high external B concentration. The kinetic assays suggest that the high B efflux and (or) exclusion are the mechanisms by which the high external B concentration in the isolated bacteria are tolerated.

Boron, in several forms, is active against several wood decay fungi (Smith 1970; Schultz et al. 1992; Cookson and Pham 1995; Kartal et al. 2004), and therefore it is used in the timber industry to protect wood from termites and fungi (Kartal et al. 2004) and in forestry to prevent infection of conifers by *Heterobasidion annosum* (Fr.) Bref. (=*Fomes annosus* (Fr.) Karst.) (Smith 1970; Schultz et al. 1992).

Eutypa dieback is a perennial canker disease of grapevines and is caused by *Eutypa lata*. This fungus produces ascospores that infect grapevines through pruning wounds during the dormant season. Fungicide applications applied during the dormant period are known to control plant fungal diseases (Irelan et al. 1999). Rolshausen and Gubler (2005) performed a field trial to show the efficacy of boric acid treatment to control infections in pruning wounds of *Eutypa lata* in a field trial. Results were that if applied at 17.5% a.i. of boron, boric acid will control this disease. The EC_{50} values for inhibiting mycelial growth and ascospore germination were 125 and 475 µg for boric acid per ml (22 and 83 µg a.i./mL), respectively. Rolshausen and Gubler (2005) also tested two 5% boric acid treatments for the control of this canker disease in different formulations: The first (a paste) was called biopaste (8.75 mg a.i./mL) and the second treatment was called bioshield, formulated as a spore suspension of *Cladosporium herbarum*. Compared to a water control both products significantly reduced the disease incidence, both in in vitro trials and in field trials. Boron did not accumulate in leaves and shoots. But, bud failure at the first node below the treated wound occurred in B-treated plants at a higher rate than that existed in untreated vines.

3 The Effect of Boron on Enzymes and Minerals

3.1 Enzymes

Boron affects the activities of at least 26 different enzymes, most of which are necessary for energy substrate metabolism (Hunt 1998). Dugger (1983) reported in an in vitro study that boron interacted with many enzymes. Boron competitively inhibited at least two classes of enzymes, one of which was oxidoreductase, an enzyme related to pyridine and flavin nucleotides. Borate also inhibited nicotin-amide adenine dinucleotide (NAD) (Roush and Norris 1950; Strittmatter 1964; Deitrich 1967; Deal 1969). Moreover, it is known that the hydroxyl groups of NAD forms complexes with the borate compound (Johnson and Smith 1976). Boron hav-ing adjacent hydroxyl groups (transferases) has a tendency to form complexes with organic molecules. It may have an interaction with important biological substances, containing polysaccharides, pyridoxine, riboflavin, dehydroascorbic acid, and the pyridine nucleotides. (Samman et al. 1998; Deviran and Volpe 2003) It binds strongly to furanoid *cis*-diols, which include erythritan, ribose, and apiose; apiose is present throughout the cell walls of vascularplants (Loomis and Durst 1992; Hunt 2012).

Nicotinamide adenine dinucleotide (NAD+) and nicotinamide adenine dinucle-otide phosphate (NADP) contain ribose components that are active in energy metabolism; binding to them affects certain metabolic pathway processes (Hunt 2012). Hunt (2012) reported that NAD+ is an essential cofactor for five sub-sub-classes of oxidoreductase enzymes and has a strong relevance for boron. The di-adenosine-phosphates (Ap_nA) are structurally similar to NAD+. Compared to NAD+, boron binding by Ap_4A, Ap_5A, and Ap6A is greatly enhanced; however, the binding is still less than that to S-adenosylmethionine (SAM).The adenine moieties of Ap_nA are driven together by hydrophobic forces and clump interfacially. Stacking of the terminal adenine moieties brings their adjacent ribose moieties into close proximity, which is the phenomenon that apparently potentiates the cooperative boron binding between opposing riboses (Hunt 2012). Boron's distinctive chemis-try allows it to react with many other metabolites and enzymes and thus may be capable of modifying mineral and energy metabolism in humans and animals (Deviran and Volpe 2003).

Boron may have a major role in controlling certain pathways that use serine pro-teases (hydrolases) or oxidoreductases (Hunt 1998, 2012). These enzymes require pyridine or flavin nucleotides (NAD+, NADP, or FAD), and by forming transition state analogs or competing for NAD or FAD, boron reversibly inhibits their activity (Hunt 1998, 2012). Serine proteases, such as thrombin, functions in regulating blood and coagulation systems. Other enzymes such as phosphoglucomutase, λ-glutamyl transpeptidase (GGT), and glyceraldehyde- 3-phosphate dehydrogenase (GPD) are also inhibited by boron (Hunt 1998; Deviran and Volpe 2003).

Boron is known to inhibit certain other enzymes (viz., aldehyde dehydrogenase, xanthine oxidase) that exist in energy metabolic pathways (Hunt 1994, 2012). Hall et al. (1989) reported that when B was orally administrated (8 mg/kg/day) to rats, daily for 14 days, LDL cholesterol and TG levels decreased. LDL bonding and LDL entrance into liver cells were also decreased, whereas fibroblasts and aorta cells showed increased HDL bonding and accumulation in liver cells. These effects were claimed to be beneficial for atherosclerosis, because they may remove cholesterol from tissues and decrease lipid accumulation (Deviran and Volpe 2003). Naghii and Samman (1997a, b) stated that boric acid intake decreased total cholesterol, HDL_3, TG, and total HDL, when given to rats for 2 weeks at a dose of 2 mg/day. However,

Green and Ferrando (1994) did not observe differences in plasma lipid concentrations, oxidation rates, LDL, or HDL fraction dispersions after 4 weeks of boric acid exposure in humans.

Ince et al. (2010) reported that boron administered to rats in the diet increased blood glutathione (GSH) concentrations and plasma vitamin C levels. In the same study, boron added to the diet at doses of 100 mg/kg increased antioxidant defense mechanisms and vitamin status of the group; neither differences between oxidant and antioxidant balance nor changes in biochemical parameters, other than serum vitamin A and liver GSH concentrations, were detected.

Another study was performed on 36 Angus and Angus-Simmental cattle that were divided into three groups. A control group received no supplementary B in their diet, a second group was fed a diet containing a 5 mg/kg supplement of B, and the third group was fed a 15 mg/kg supplement of B for 47 days to determine the effect on disease resistance to bovine herpesvirus type-1 (BHV-1). The cattle were inoculated with BHV-1 intranasally on the 34th day. On the second day following inoculation, rectal temperatures of the cattle and plasma tumor necrosis factor-α concentrations had increased ($P<0.05$). On the fourth day after inoculation, the plasma acute phase proteins had multiplied ($P<0.01$), and plasma interferon-γ levels were beginning to decline ($P<0.05$). Plasma B concentrations had increased slightly after the addition of B ($P<0.001$), whereas the dietary levels of B fed showed no significant effect on BHV-1 symptoms, and had little influence on plasma acute phase proteins and cytokines (Fry et al. 2010).

3.2 Minerals

Boron has a regulatory role in the metabolism of several minerals such as phosphorous, magnesium, calcium, and molybdenum (Wilson and Ruszler 1996). Hunt et al. (1983) showed that chicken growth was enhanced when they were exposed to boron–calcium and boron–magnesium and suggested that the relationship between magnesium and boron was stronger than that between calcium or phosphorus and boron. A relationship appeared when the boron: magnesium molar ratio was quite low in both plasma and in the consumed diet, although a direct effect of boron on magnesium metabolism was not thought to have occurred. Apparently, boron indirectly influences magnesium metabolism, and ultimately, calcium and phosphorus metabolism by influencing an enzyme or the hormone system (Hunt et al. 1983).

The regulatory role that B plays with minerals affects bone health and is particularly interesting, because it affects the relationship B has with magnesium, vitamin D, phosphorus, and calcium (Deviran and Volpe 2003; Nielsen 1990). A deficiency of boron upsets the intrinsic plasma concentration balance between calcium, magnesium, and phosphorus (Hegsted et al. 1991). Hunt and Nielsen (1986) studied magnesium-deficient chicks and found that supplemental boron decreased the incidence of abnormalities caused by insufficient magnesium intake. Subsequent boron supplementation enhanced growth and increased plasma calcium and magnesium concentrations, and also inhibited the calcification of cartilage (Hunt 1989).

The effect that vitamin D_3 has on glycolysis may be related to calcium levels, because calcium is the main inhibitor of phosphofructokinases (limiting enzymes in glycolytic paths) (Auffermann et al. 1990). In the cartilage of rachitic rats, the glycolysis rate more than doubles, commensurate with increased activities of the phosphofructokinases, aldolase, pyruvate kinase, and lactate dehydrogenase (Meyer and Kunin 1969). Similarly, in chronic kidney failure, glucose tolerance and hyperlipoproteinemia occur, and after synthetic vitamin D_3 treatment starts, fasting blood glucose and TG levels decrease and glucose tolerance is reduced (Lind et al. 1988). Researchers have also indicted that vitamin D_3 is essential for insulin secretion (Norman et al. 1980; Gedik and Akalin 1986). A dietary vitamin D_3 deficiency reduces hepatic glycogen in rats (Davis et al. 1989). Unlike the hypothesis that is worded "lipid–carbohydrate is not transformed in the livers of mammals", glycogen content was claimed to have increased 35% after the incubation of rat liver sections to which vitamin D_3 with palmitate was added (Davis et al. 1989). Hunt et al. (1994) reported that boron modulated hepatic glycolysis when a vitamin D3 deficiency existed; moreover, when boron was added to the diet (2.25 vs. 0.16 mg/kg) it reduced effects on glycolytic metabolites such as fructose-1,6-diphosphate P_2, glycerate-2P, and $(OH)_2$-acetone P in freeze-clamped chick liver (Hunt 1989). Supplemental boron in the diet reduced the plasma pyruvate concentrations in vitamin D_3-deprived rats (Hunt et al. 1994). These data suggest that boron limits the activity of some enzymes and stabilizes reactive compounds by regulating energy substrate utilization (Hunt et al. 1994).

4   Boron Toxicity

Micronutrient elements may be toxic at some dose, duration of exposure time, and application method (Blevins and Lukaszewski 1994). The toxicity of borate compounds have been extensively studied in both laboratory and other animals. Boric acid and borax were the forms of boron most commonly administered to animals in such testing. Boric acid and borax have performed toxicologically similarly in the species to which they have been administered (World Health Organization 1998). In boron exposure studies, whether borax or boric acid was tested, data are expressed as boron equivalents to enable data comparisons (USDA Forest Service 2006). At a physiologic pH, borate salts are converted almost entirely to nonionized boric acid; hence, boric acid and borate salts have similar toxicologic features (USDA Forest Service 2006). Following oral administration, inorganic borates are well absorbed (about 90% of the administered dose) by animals (USDA Forest Service 2006).

Animal experiments revealed that toxicity results from dietary boron intake that exceeds about 100 µg g^{-1}. Testicular cell damage and atrophy may appear when dietary boron levels in mice exceed 4,000 mg boric acid kg^{-1}. Similar effects were shown for rats after oral administration (Nielsen 1997; EPA 2008). It was reported that 17.5 mg B/kg per day (blood boron level of ~2,020 ng/g) affected fertility, and 9.6 mg B/kg per day (blood boron level of ~1,270 ng/g) affected the normal development of exposed rats (Price et al. 1997; Scialli et al. 2010).

In humans, acute toxicity symptoms of excessive boron exposure are nausea, vomiting, diarrhea, and lethargy (Linden et al. 1986; Nielsen 1997). Oral intake by two infants from dipping pacifiers into a preparation of borax and honey resulted in scanty hair, anemia, seizures, and patchy dry erythema, over a period of several weeks (Gordon et al. 1973; Nielsen 1997). Signs of chronic boron toxicity are poor appetite, nausea, weight loss, and decreased sexual activity, seminal volume, sperm count, and motility (Nielsen 1997; USDA Forest Service 2006). However, Duydu et al. (2012) pointed out that mean blood boron levels of boric acid workers were 223.89 ng/g, and this level was ~9 times lower than levels that produced reproductive effects in exposed rats; these levels were also ~6 times lower than levels that produced developmental effects in rats. At the levels recorded for exposed workers, no harmful effect on the reproductive system was observed. In addition, urine analyses indicated that the boron exposure of boron mine workers was 6.5 mg/day, and these levels produced no negative effects on worker semen profiles (Korkmaz et al. 2011). Workers who were exposed to the highest boron levels (i.e., 125 mg B/day) showed no significant effects on semen characteristics (Scialli et al. 2010). In conclusion, evidence suggests that the daily exposure experienced by humans to boric acid or sodium borates is unlikely to produce reproductive toxicity (Duydu et al. 2012).

An acute oral dose of borax produced an LD_{50} of 4.50 g/kg in rats and boric acid an LD_{50} of 3.45 g/kg by gavage; after dosing, rats displayed depression, ataxia, convulsions, and death (Weir and Fisher 1972). Sabuncuoglu et al. (2006) reported that a subacute dose of 400 mg/kg/day administered orally to rats produced histopathological changes in kidney tissue. A single oral dose of borax administered to dogs by capsule (at levels of 1.54–6.51 g borax/kg or 0.174–0.736 g B/kg) or boric acid administered by capsule (1.0–3.98 g boric acid/kg or 0.175–0.697 g B/kg) caused no dog deaths (Weir and Fisher 1972). The acute oral lethal dose of boric acid in 1-day-old chickens was found to be 2.95 ± 0.35 g/kg. One-day-old broiler chicks were housed in floor pens, in which the litter had been treated with 0, 0.9, 3.6, or 7.2 kg of boric acid per 9.9 m^2 of floor space. No B residue level elevation was seen in brain, kidney, liver, or white muscles of the chicks. *Ad libitum* feeding of boric acid at 500 ppm or 1,250 ppm to chicks did not alter boron levels in tissue. However, doses of 2,500 ppm or 5,000 ppm boric acid to the chicks raised residue tissue levels. Hence, it was realized that broilers grown on boric acid-treated litter do not consume enough boric acid to elevate boron levels above the norm in their tissues (viz., brain, kidney, liver, and white muscle) (Sander et al. 1991).

Restuccio et al. (1992) reported that an acute oral dose of boron in a human produced toxicity and death. A 45-year-old man drank two cups of boric acid crystals that were dissolved in water with the aim to commit suicide; the man died on the third day after consuming the boron crystals (Restuccio et al. 1992). The symptoms noted soon after consuming this dose ware nausea, vomiting, greenish diarrhea, and dehydration (Restuccio et al. 1992). Hypotension, metabolic acidosis, oliguric renal failure, a generalized erythematous rash, and several superficial skin abrasions presented within 2 days of the exposure (Restuccio et al. 1992). The condition of the poisoned man failed to improve despite injecting intravenous fluids and vasopressors. Atrial fibrillation with a rapid ventricular response was produced and sinus rhythm could not be restored (Restuccio et al. 1992). This rhythm deteriorated to electromechanical dissociation, and the patient died 17 h after hospital admission (Restuccio et al. 1992). The urine and whole blood boric acid concentrations, approximately 52 h after ingestion, were 160 and 42 mg/dL, respectively (Restuccio et al. 1992). These results are equivalent to urine and blood boron concentrations of 28 and 7 mg/dL, respectively. The postmortem urine boron concentration was 29.4 mg/dL (Restuccio et al. 1992).

The subchronic oral administration of 1 g/kg borax and boric acid for 1–3 weeks rats produced a decrease in body weight, DNA synthesis inhibition, and clinical toxicity signs after 3 weeks (Dani et al. 1971). A 90-day test in dogs fed dietary borax at concentrations of 0%, 0.0154%, 0.154%, and 1.54% showed no treatment-related effect for any blood or urine value in males. In female dogs, hematocrit and hemoglobin decreased in the 1.54% treatment group; similarly, in the male 1.54% treatment group, testicular weights were decreased, testicular atrophy occurred, and alterations were detected in the seminiferous tubules (Paynter 1963).

A chronic oral study with borax and boric acid in rats was performed at dietary concentrations of 117, 350, and 1,170 ppm boron equivalents. Results were that no harmful effects were produced at the lower intake levels, but at the high dose level (1,170 ppm) clinical signs of toxicity occurred that included coarse hair, scaly tails, hunched posture, swelling and desquamation of the pads of paws, shrunken scrotum in males, testicular atrophy, decreased testes weight, atrophied seminiferous epithelium and decreased tubular size, as well as inflamed eyes with a bloody discharge, and a decrease in packed cell volume and hemoglobin of blood (Weir and Fisher 1972). No evidence of carcinogenesis was observed in any treatment group (Weir and Fisher 1972). Dogs fed with 1,170–2,000 ppm boron for 2 years showed stunted growth, less productive food use, skin rashes, and gonadal deterioration (Weir and Fisher 1972).

The dermal effects of boron application were investigated on ten New Zealand Rabbits by applying borax (sodium tetraborate decahydrate) as a single dose to clipped skin at 2.0 g/kg ($LD_{50} > 2.0$ g/kg) and then occluding the applied material for 2 h (Reagan 1985a). The symptoms observed on the treated animals included anorexia, decreased activity, diarrhea, soft stools, and nasal discharge (Reagan 1985a). The borax (sodium tetraborate decahydrate) application on shaved skin at 0.5 g/kg caused no skin irritation (Reagan 1985b).

Borax (sodium tetraborate decahydrate), at a dose of 0.1 g, was instilled into the eyes of six New Zealand Rabbits, caused severe irritation to the iris, and produced corneal opacity and conjunctival redness, chemosis, and discharge (Reagan 1985c). In another study, a 4 h daily inhalation of 2.0 mg/L borax (sodium tetraborate deca-hydrate) over 14 days in rats caused no mortality (Wnorowski 1994).

Researchers explored if borax could serve as an antidote, and, indeed, it was found to be a beneficial antagonist for aluminum toxicity (Turkez et al. 2012).

The borax (3.25 and 13 mg/kg bwt) clearly protected exposed rats against aluminum (AlCl$_3$; 5 mg/kg bwt)-induced toxicity (Turkez et al. 2012).In particular, borax blocked the increase of micronucleated hepatocytes and significantly modulated the genotoxic effects caused by AlCl$_3$ (Turkez et al. 2012).

Boron is toxic to particular processes in vascular plants. Some of the effects noted in plants included the following: altered plant metabolism, lowered cell division in roots, reduced leaf chlorophyll content and photosynthetic ratio, and reduced lignin and suberin levels, among others (Nable et al. 1997; Reid 2007). Therefore, decreased shoot and root growth is typical of plants exposed to high B levels (Nable et al. 1990). The distribution of B in leaves follows a particular pattern that extends from leaf base to tip in some plants, and this leads to particular toxicity symptoms on older leaves, which may show tip chlorosis or necrosis, or both (Marschner 1995; Roessner et al. 2006; Camacho-Cristóbal et al. 2008). Considering the chemical characteristics of B, it has been proposed that B-induced plant toxicity derives from three main causes: (a) changes to cell wall structure; (b) metabolic disorder from binding to ribose moieties in certain molecules (viz., adenosine triphosphate (ATP), nicotinamide adenine dinucleotide (reduced form) (NADH), or nicotinamide adenine dinucleotide phosphate (reduced form) (NADPH)); and/or (c) disrupting cell division and development by binding to ribose, either the free sugar or ribose within RNA (Reid et al. 2004). Reid et al. (2004) reported that none of these proposed mechanisms have been confirmed, and further, there is no proof to support another hypothesis that leaf toxicity results from osmotic stress induced by B accumulation.

Konuk et al. (2007) investigated the effect of boron on the mitotic index of *Allium cepa* root meristematic cells. In this study, the rate of growth inhibition was first determined, and then different concentrations of boron were tested to determine effects on onion tuber roots. Distilled water was used as the control. Because the *Allium cepa* cell cycle is 24 h, B was applied at 12, 24, and 48 h. For each dose, the mitotic index and mitotic phase frequencies were calculated separately. Most abnormalities were observed to occur at c-metaphase, prometaphase, and disturbed anaphase–telophases. In addition, effects were observed on the anaphase bridge, on polyploidy, and on late chromosome particles.

5   Conclusions

Boron is thought to be an essential element for animals, people, and plants. Although several studies have been performed to determine the effects of boron on fertility and general health of animals, and human data are also available, the overall picture of how safe boron may be is still incomplete. In particular, we conclude that:

1. Boron does produce effects on human and animal bone development, mineralization, Ca, P, and Mg metabolism, energy metabolism, and activates certain enzymes. Intake of boron minerals can be effective for optimizing the treatment

of bone structure disorders, can reduce cholesterol, can improve lipid metabolic profiles, and can reduce triglyceride levels in humans and animals. In addition, sub-toxic doses of boron may be beneficial for preventing and treating type 2 diabetes and alcohol-induced fatty liver.

2. Humans and animals who reside near areas where boron is mined are exposed to this element. Therefore, research is needed both to determine what optimum levels of boron intake should be for animals and humans, and what exposure limits, if any, are needed for boron in mining regions.

3. In periparturient dairy cattle, boron intake may be particularly important for preventing or treating fatty liver, hypocalcemia (milk fever), and hypomagnesemia.

4. Boron may affect the yield and growth performance of fruits, vegetables, nuts, and grains. Although waterless borax and borax pentahydrate are used as fertilizers, their effects on plants are unclear. In addition, the effects of boron on humans and animals who consume plants or plant products to which B fertilizer was applied are unknown.

5. The effects on soil, water, or wildlife of the boron used in agriculture as both pesticide and fertilizer are unknown.

Additional research is needed to address the gaps in knowledge that were indicated above. In addition, research is needed to determine what dose of boron or boron compounds are optimal in animal rations. Finally, future research should be undertaken to enhance the understanding of how boron affects metabolic systems, and to explain how boron functions mechanistically.

6 Summary

Boron is a mineral used in human and animal health, in agriculture as a fertilizer and pesticide, and in the manufacturing of several commodities (e.g., glass, ceramics, automotive components, paint, etc.). Humans and animals consume boron daily via dietary intake. However, what the daily intake level is remains unclear. Recently, researchers have concluded that boron intake is essential for plants, animals, and humans, although insights as to what this element's biological effects are is still limited.

Although several studies on the effects of boron and how it functions have been performed over the last decade, more information is needed to clarify both its effects and how it produces its action. It is clear that boron exposure affects many enzymes and enzyme systems; some of the symptoms observed in animals from exposure to boron are explained by enzyme effects. Boron is known to produce effects on fat and lipid metabolism, on minerals and mineral metabolism, and on vitamin D. In addition, boron affects bone development. In addition, the effects of several different forms of boron on poultry and laboratory animals have been determined.

Although much of the information on boron that has already been collected is useful, much more is needed. Prospectively, the result of current research suggests that boron may be useful in the future for preventing obesity, fatty liver, and diabetes, or may be used to prevent or treat bone health problems in both humans and domestic animals.

References

Ahmed I, Fujiwara T (2010) Mechanism of boron tolerance in soil bacteria. Canadian J Microbiol 56(1):22–26

Anderson DL, Cunningham WC, Lindstrom TR (1994) Concentrations and intakes of H, B, S, K, Na, Cl, and NaCl in foods. J Food Comp Anal 7:59–82

Argust P (1998) Distribution of boron in the environment. Biol Tr Elem Res 66:131–143

Armstrong TA, Spears JW, Crenshaw TD, Nielsen FH (2000) Boron supplementation of a semipurified diet for weanling pigs improves feed efficiency and bone strength characteristics and alters plasma lipid metabolites. J Nutr 130:2575–2581

Auffermann W, Wagner S, Wu S, Buser P, Parmley W, Wikman-Coffelt J (1990) Calcium inhibition of glycolysis contributes to ischaemic injury. Cardio vasc Res 24:510–520

Bai Y, Hunt CD (1996) Dietary boron enhances efficacy of cholecalciferol in broiler chicks. J Trace Elem Exp Med 9:117–132

Baker SJ, Ding CZ, Akama T, Zhang YK, Hernandez V, Xia Y (2009) Therapeutic potential of boron-containing compounds. Future Med Chem 1(7):1275–1288

Bakken NA, Hunt CD (2003) Dietary boron decreases peak pancreatic in situ insulin release in chicks and plasma insulin concentrations in rats regardless of vitamin D or magnesium status. J Nutr Nov 133(11):3577–3583

Basoglu A, Sevinc M, Guzelbektas H, Civelek T (2000) Effect of borax on lipid profile in dogs. Online J Vet Res 4(6):153–156

Basoglu A, Sevinç M, Birdane FM, Boydak M (2002) Efficacy of sodium borate in the prevention of fatty liver in dairy cows. J Vet Intern Med 16:732–735

Basoglu A, Baspinar N, Ozturk AS, Akalin PP (2010) Effects of boron administration on hepatic steatosis, hematological and biochemical profiles in obese rabbits. Trace Elements and Electrolytes 27:225–231

Basoglu A, Baspinar N, Ozturk AS, Akalin PP (2011) Effects of long-term boron administration on high-energy diet-induced obesity in rabbits: NMR-based metabonomic evaluation. J Anim and Veterinary Adv 10(12):1512–1515

Blevins DG, Lukaszewski KM (1994) Proposed physiologic functions of boron in plants pertinent to animal and human metabolism. Environ Health Perspect 102(Suppl 7):31–33

Blevins DG, Lukaszewski KM (1998) Boron in plant structure and function. Annu Rev Plant Physiol Plant Mol Biol 49:481–500

Bobe G, Young JW, Beitz DC (2004) Pathology, etiology, prevention, and treatment of fatty liver in dairy cows. J Dairy Sci 87:3105–3124

Bolaños L, Lukaszewski K, Bonilla I, Blevins D (2004) Why boron? Plant Physiol Biochem 42:907–912

Camacho-Cristóbal JJ, Rexach J, Fontes AG (2008) Boron in plants: deficiency and toxicity. Journal of Integrative Plant Biolo 50(10):1247–1255

Clark WB, Koekebakker M, Barr RD, Dowining RG, Fleming RF (1987) Analysis of ultratrace lithium and boron by neutron activation and mass-spectrometric measurement of 3He and 4He. Appl Radiat Isot 38:735–743

Cookson LJ, Pham K (1995) Relative tolerance of twenty Basidiomycetes to boric acid. Material Organismen 29:187–196

Dani HM, Saini HS, Allag IS, Singh B, Sareen K (1971) Effect of boron toxicity on protein andnucleic acid contents of rat tissues. Res Bull Panjab Univ Sci 22(1–2):229–235

Davis W, Matthews J, Goodman D (1989) Glyoxylate cycle in the rat liver: effect of vitamin D_3 treatment. FASEB J 3:1651–1655

Deal WJ (1969) Metabolic control and structure of glycolytic enzymes IV. Nicotinamide-adenine dinucleotide dependent in vitro reversal of dissociation and possible in vivo control of yeast glyceraldehyde 3-phosphate dehydrogenase synthesis. Biochemistry 8:2795–2805

Deitrich R (1967) Diphosphopyridine nucleotide-linked aldehyde dehydrogenase III. Sulfhydryl characteristics of the enzyme. Arch Biochem Biophys 119:253–263

DeLuca H, Schnoes H (1983) Vitamin D: recent advances. Ann Rev Biochem 52:411–439

Deviran TA, Volpe SL (2003) The physiological effects of dietary boron. Crit Rev Food Sci and Nutrit 43(2):219–231

Dugger W (1983) Boron in plant metabolism. In: Lauchli A, Bielesk R (eds) Encyclopedia of plant physiology. Inorganic plant nutrition, vol 15. Springer-Verlag, Berlin, pp 626–650

Duydu Y, Basaran N, Bolt HM (2012) Exposure assessment of boron in Bandırma boric acid production plant. J Trace Elem Med Biol 26(2–3):161–164

Eckhert CD (1998) Boron stimulates embryonic trout growth. J Nutr 128:2488–2493

EPA (U.S. Environmental Protection Agency) (2008) Health effects support document for boron. EPA document number EPA-822-R-08-002 January

Fort DJ, Stover EL, Strong PL, Murray FJ, Keen CL (1999) Chronic feeding of a low boron diet adversely affects reproduction and development in Xenopus laevis. J Nutr 129:2055–2060

Fry RS, Lloyd KE, Jacobi SK, Siciliano PD, Robarge WP, Spears JW (2010) Effect of dietary boron on immune function in growing beef steers. J Anim Physiol AnimNutrit 94:273–279

Gedik O, Akalin S (1986) Effects of vitamin D deficiency and repletion on insulin and glucagon secretion in man. Diabetologia 29:142–145

Goldbach HE (1997) A critical review on current hypotheses concerning the role of boron in higher plants: suggestions for further research and methodological requirements. J Trace Microprobe Tech 15:51–91

Gordon AS, Prichard JS, Freedman MH (1973) Seizure disorders and anaemia associated with chronic borax intoxication. Can Med Assoc J 108:719–721

Green NR, Ferrando AA (1994) Plasma boron and the effects of boron supplementation in males. Environ Health Perspect 102:73–77

Hall LH, Spielvogal BF, Griffin TS, Docks EL, Brotherton RJ (1989) The effects of boron hyperlipidemic agents on LDL and HDL receptor binding and related enzyme activities of rat hepatocytes, aorta cells and human fibroblasts. Res Comm Chem Pathol Pharmocol 65:297–317

Hamilton EI, Minsky MJ, Cleary JJ (1972) The concentration and distribution of some stable elements in healthy human tissues from the United Kingdom. Sci Tot Env 1:341–374

Hegsted M, Keenan MJ, Siver F, Wozniak P (1991) Effect of boron on vitamin D deficient rats. Biol Trace Elem Res 28:243–256

Howe PD (1998) A review of boron effects in the environment. Biol Tr Elem Res 66:153–166

Hunt CD (1989) Dietary boron modified the effects of magnesium and molybdenum on mineral metabolism in the cholecalciferol-deficient chick. Biol Trace Elem Res 22:201–220

Hunt CD (1994) The Biochemical effects of physiologic amounts of dietary boron in animal nutrition models. Environl Health Perspect 102(7):35–42

Hunt CD (1998) One possible role of dietary boron in higher animals and humans. Biol Trace Elem Res 66:205–225

Hunt CD (2003) Dietary boron: an overview of the evidence for its role in immune function[†]J. Trace Elem Exp Med 16:291–306

Hunt CD (2012) Dietary boron: progress in establishing essential roles in human physiology. J Trace Elements in Med and Biol 26:157–160

Hunt CD, Herbel JL (1991–1992) Boron affects energy metabolism in the streptozotocin-injected, vitamin D3-deprived rat. Magnesium Trace Elem 10:374–386

Hunt CD, Herbel JL (1993) Physiological amounts of dietary boron improve growth and indicators of physiological status over a 20-fold range in the vitamin D3-deficient chick. In: Anke M,

Meissner D, Mills C (eds) Trace element metabolism in man and animals, vol 8. Verlag Media Touristik, Gersdorf, Germany, pp 714–718

Hunt CD, Nielsen F (1981) Interaction between boron and cholecalciferol in the chick. In: Gawthorne J, White C (eds) Trace element metabolism in man and animals, vol 4. Australian Academy of Science, Canberra, pp 597–600

Hunt CD, Nielsen FH (1986) Dietary boron affects bone calcification in magnesium- and cholecalciferoldeficient chicks. In: Underwood EJ (ed) Trace elements in human and animal nutrition, 5th edn. Academic, New York, pp 275–277

Hunt CD, Shuler T, Nielsen F (1983) Effect of boron on growth and mineral metabolism. In: Anke M, Baluman W, Braunlich H, Bruckner C (eds) 4 Spurenelement-symposium. Friedrich-Schiller University, Jena, Germany, pp 149–155

Hunt CD, Halas E, Eberhardt M (1988) Long-term effects of lactational zinc deficiency on bone mineral composition in rats fed a commercially modified Luecke diet. Biol Trace Elem Res 16:97–113

Hunt CD, Shuler T, Mullen L (1991) Concentration of boron and other elements in human foods and personal-care products. J Am Diet Assoc 91:558–568

Hunt CD, Herbel JL, Idso JP (1994) Dietary boron modifies the effects of vitamin D_3 nutriture on indices of energy substrate utilization and mineral metabolism in the chick. J Bone Miner Res 9:171–181

Ince S, Kucukkurt I, Cigerci IH, Fatih Fidan A, Eryavuz A (2010) The effects of dietary boric acid and borax supplementation on lipid peroxidation, antioxidant activity, and DNA damage in rats. J Trace Elem Med Biol Jul 24(3):161–164

Indraprasit S, Alexander GV, Gonick HC (1974) Tissue composition of major and trace elements in uremia and hypertension. J Chron Dis 27:135–161

Irelan N, Gubler WD, DeScenzo R (1999) Efficacy testing of Eutypa chemical and biological control candidates with DNA-based diagnostics. Winegrowing January-February:47–56

Iyengar GV, Clarke WB, Downing RG (1990) Determination of boron and lithium in diverse biological matrices using neutron activation-mass spectrometry (NA-MS). Fres J Anal Chem 338:562–566

Johnson S, Smith K (1976) The interaction of borate and sulfite with pyridine nucleotides. Biochemistry 15:553–559

Kabu M, Civelek T (2012) Effects of propylene glycol, methionine and sodium borate on metabolic profile in dairy cattle during periparturient period. Revue Méd Vét 163(8–9):419–430

Kartal SN, Yoshimura T, Imamura Y (2004) Decay and termite resistance of borontreated and chemically modified wood by in situ co-polymerisation of allyl glycidyl ether (AGE) with methyl methacrylate (MMA). Int Biodeterioration Biodegradation 53:111–117

Kılıç AM, Kılıç Ö, Andaç İ, Çelik AG (2009) Boron mining in Turkey, the marketing situation and the economical importance of Boron in the World IV, International boron symposium, Eskişehir-TURKEY, 15–17 Oct

Konuk M, Liman R, Cigerci IH (2007) Determination of genotoxic effect of boron on Allium Cepa root meristematic cells. Pak J Bot 39(1):73–79

Korkmaz M, Yenigün M, Bakırdere S, Ataman OY, Keskin S, Müezzinoğlu T, Lekili M (2011) Effects of chronic boron exposure on semen profile. Biol Trace Elem Res 143(2):738–750

Lind L, Lithell H, Wengle B, Wrege U, Ljunghall S (1988) A pilot study of metabolic effects of intravenously given alpha-calcidol in patients with chronic renal failure. Scand J Urol Nephrol 22:219–222

Linden CH, Hall AH, Kulig KW, Rumack BH (1986) Acute ingestions of boric acid. Clin Toxicol 24:269–279

Loomis WD, Durst RW (1992) Chemistry and biology of boron. Biofactors 3:229–239

Lovatt C, Dugger W (1984) Boron. In: Frieden E (ed) Biochemistry of the essential trace elements. Plenum, New York, pp 389–421

Marschner H (1995) Mineral nutrition of higher plants, 2nd edn. Academic Press, San Diego, pp 379–396

Massie HR, Whitney SJ, Aiello VR, Sternick SM (1990) Changes in boron concentration during development and aging of Drosophilia and effect of dietary boron on life span. Mech Aging Dev 53:1–7

Meyer W, Kunin A (1969) The inductive effect of rickets on glycolytic enzymes of rat epiphyseal cartilage and its reversal by vitamin D and phosphate. Arch Biochem Biophys 129:438–446

Nable RO, Cartwright B, Lance RC (1990) Genotypic differences in boron accumulation in barley: relative susceptibilities to boron deficiency and toxicity. In: El Bassam N, Dambroth M, Loughman B (eds) Genetic aspects of plant mineral nutrition. Kluwer Academic Publishers, Dordrecht, The Netherlands, pp 243–251

Nable RO, Ba˜ nuelos GS, Paull JG (1997) Boron toxicity. Plant Soil 193:181–198

Naghii MR (1999) The significance of dietary boron, with particular reference to athletes. Nutr Health 13:31–37

Naghii MR, Samman S (1997a) The effect of boron supplementation on its urinary excretion and selected cardiovascular risk factors in healthy male subjects. Biol Tr Elem Res 56:273–286

Naghii MR, Samman S (1997b) The effect of boron on plasma testosterone and plasma lipids in rats. Nutr Research 17:523–531

Naghii MR, Mofid M, Asgari AR, Hedayati M, Daneshpour MS (2011) Comparative effects of daily and weekly boron supplementation on plasma steroid hormones and proinflammatory cytokines. J Trace Elem Med Biol 25(1):54–58

Newnham RE (1977) Mineral imbalance and boron deficiency. In: Underwood EJ (ed) Trace elements in human and animal nutrition, 4th edn. Academic Press, Inc., New York, pp 400–402

Newnham RE (1991) Agriculture practices affect arthritis. Nutr Health 7:89–100

Nielsen FH (1988) Boron- an overlooked element of potential nutritional importance. Nutr Today Jan/Feb:4–7

Nielsen FH (1990) Studies on the relationship between boron and magnesium which possibly affects the formation and maintenance of bones. Magnes Trace Elem 9:61–69

Nielsen FH (1997) Boron in human and animal nutrition. Plant Soil 193:199–208

Nielsen FH, Hunt CD, Mullen LM, Hunt JR (1987) Effect of dietary boron on mineral, estrogen, and testosterone metabolism in postmenopausal women. FASEB J 1(5):394–397

Nielsen FH, Shuler TR, Zimmerman TJ, Uthus EO (1988) Magnesium and methionine deprivation affect the response of rats to boron deprivation. Biol Trace Elem Res 17:91–107

Norman A, Heldt A, Grodsky G (1980) Vitamin D deficiency inhibits pancreatic secretion of insulin. Science 209:823–825

Pawa S, Ali S (2006) Boron ameliorates fulminant hepatic failure by counteracting the changes associated with the oxidative stress. Chem Biol Interact 160(2):89–98

Paynter OE (1963) 90-Day dietary feeding - Dogs with 20 MULE TEAM® Borax (Sodium tetraborate decahydrate). MRID 406923–07

Penland JG (1998) The importance of boron nutrition for brain and psychological function. Biol Tr Elem Res 66:299–317

Price CJ, Strong PL, Murray FJ, Goldberg MM (1997) Blood boron concentration in pregnant rats fed boric acid throughout gestation. Reprod Toxicol 11(6):833–842

Reagan E (1985) Acute dermal toxicity study of 20 MULE TEAM sodium tetraboratedecahydrate in New Zealand white rabbits: lab project number: 8403A. Unpublished study prepared by Food & Drug Research Labs, Inc. 9 p. MRID 43553200

Reagan E (1985) Primary dermal irritation study of 20 MULE TEAM sodium tetraboratedecahydrate in New Zealand white rabbits: lab project number: 8403B. Unpublished study prepared by Food & Drug Research Labs, Inc. 8 p. MRID 43553201

Reagan E (1985) Primary eye irritation study of 20 MULE TEAM sodium tetraborate decahydrate in New Zealand white rabbits: lab project number: 8403B. Unpublished study prepared by Food & Drug Research Labs, Inc. 21 p. MRID 43553202

Reid R (2007) Update on boron toxicity and tolerance in plants. In: Xu F, Goldbach HE, Brown PH, Bell RW, Fujiwara T, Hunt CD, Goldberg S, Shi L (eds) Advances in plant and animal boron nutrition. Springer, Dordrecht, The Netherlands, pp 83–90

Reid RJ, Hayes JE, Post A, Stangoulis JCR, Graham RD (2004) A critical analysis of the causes of boron toxicity in plants. Plant Cell Environ 25:1405–1414

Restuccio A, Mortensen ME, Kelley MT (1992) Fatal ingestion of boric acid in an adult. Am J Emerg Med Nov 10(6):545–547

Roessner U, Patterson JH, Forbes MG, Fincher GB, Langridge P, Bacic A (2006) An investigation of boron toxicity in barley using metabolomics. Plant Physiol 142:1087–1101

Rolshausen PE, Gubler WD (2005) Use of boron for the control of Eutypa Dieback of grapevines. Plant Dis 89:734–738

Roush A, Norris E (1950) The inhibition of xanthine oxidase by borates. Arch Biochem Biophys 29:344–347

Sabuncuoglu BT, Kocaturk PA, Yaman Ö, Kavas GO, Tekelioglu M (2006) Effects of subacute boric acid administration on rat kidney tissue. Clin Toxicol (Phila) 44(3):249–253

Samman S, Naghii MR, Lyons Wall PM, Verus AP (1998) The nutritional and metabolic effects of boron in humans and animals. Biol Tr Elem Res 66:227–235

Sander JE, Dufour L, Wyatt RD, Bush PB, Page RK (1991) Acute toxicity of boric acid and boron tissue residues after chronic exposure in broiler chickens. Avian Dis 35(4):745–749

Schultz ME, Parmeter JR, Slaughter GW (1992) Long-term effect of treating true fir stumps with sodium tetraborate to control losses from Heterobasision annosum. West J Appl For 7:29–31

Scialli AR, Bondeb JP, Irene Brüske-Hohlfeldc B, Culverd D, Li Y, Sullivanf FM (2010) An overview of male reproductive studies of boron with an emphasis on studies of highly exposed Chinese workers. Reprod Toxicol 29:10–24

Simon J, Rosselin G (1978) Effect of fasting, glucose, amino acids and food intake on in vivo insulin release in the chicken. Horm Metab Res 10:93–98

Smith RS (1970) Borax to control Fomes annosus infection of white fir stumps. Plant Dis Rep 54:872–875

Strittmatter P (1964) Reversible direct hydrogen transfer from reduced pyridine nucleotides to cytochrome b5 reductase. J Biol Chem 239:3043–3050

Sutherland B, Strong P, King JC (1998) Determining human dietary requirements for boron. Biol Tr Elem Res 66:193–204

Tamay-Cach F, Correa-Basurto J, Villa-Tanaca L, Mancilla-Percino T, Juárez-Montiel M, Trujillo-Ferrara JG (2012) Evaluation of new antimicrobial agents on Bacillus spp. strains: docking affinity and in vitro inhibition of glutamate-racemase. J Enzyme Inhib Med Chem Aug 7. [Epub ahead of print] DOI: 10.3109/14756366.2012.705837

Turkez H, Geyikoğlu F, Tatar A (2012) Borax counteracts genotoxicity of aluminum in rat liver. Toxicol Ind Health Published online 4 April 2012 DOI: 10.1177/0748233712442739

USDA Forest Service (2006) Human health and ecological risk assessment for borax (Sporax®) final report

Vanderpool RA, Johnson PE (1992) Boron isotope ratios in commercial produce and boron-10 foliar and hydroponic enriched plants. J Agric Food Chem 40:462–466

Weir RJ, Fisher RS (1972) Toxicologic studies on borax and boric acid. Toxicol Appl Pharmacol 23:351–364

WHO (World Health Organization) (1998) Environmental health criteria 204: boron. International programme on chemical safety, Geneva, Switzerland. ISBN 92 4 157204 3, pp. 105–106

Wilson JH, Ruszler PL (1996) Effects of dietary boron supplementation on laying hens. Br Poul Sci 37:723–729

Wnorowski G (1994) Sodium tetraborate decahydrate: acute inhalation toxicity limit test (in rats). Lab project number: 3309. Unpublished study prepared by Product Safety Labs. 24p. MRID 43500800

Woods WG (1994) An introduction to boron: history, sources, uses, and chemistry. Environ Health Perspect 102(Suppl 7):5–11

Environmental Fate and Ecotoxicology of Fenpropathrin

Emerson Kanawi, Robert Budd, and Ronald S. Tjeerdema

Contents

E. Kanawi (✉) • R.S. Tjeerdema
Department of Environmental Toxicology, College of Agricultural
& Environmental Sciences, University of California, Davis, CA 95616-8588, USA
e-mail: Ekanawi@cdpr.ca.gov

R. Budd
Department of Pesticide Regulation, California Environmental Protection Agency,
Sacramento, CA 95812-4015, USA

D.M. Whitacre (ed.), *Reviews of Environmental Contamination and Toxicology*,
Reviews of Environmental Contamination and Toxicology 225,
DOI 10.1007/978-1-4614-6470-9_3, © Springer Science+Business Media New York 2013

1 Introduction

Fenpropathrin ((RS)-α-cyano-3-phenoxybenzyl-2,2,3,3-tetramethylcyclopropane-1-carboxylate; Fig. 1) is a racemic mixture, broad-spectrum pyrethroid insecticide, and acaricide. Discovered by Sumitomo Chemical Company Ltd., then developed by Valent USA, fenpropathrin was the first of the light-stable pyrethroids to be produced. First synthesized in 1971 and commercialized in 1980 (Davies 1985), the technical product (90% purity) is formulated as an emulsifiable concentrate (30.9% active ingredient) and is registered in California as Danitol 2.4 EC Spray and Tame 2.4 EC Spray (CDPR 2012b). It is classified as a type II pyrethroid, characterized by the addition of a cyano group at the benzylic carbon. Such α-cyano pyrethroids have enhanced insecticidal activity because of their affinity for voltage-gated membrane channels (Bailey 2009). Synonymous chemical and common names include: α-cyano-3-phenoxybenzyl 2,2,3,3-tetramethyl-1-cyclopropanecarboxylate; Danitol; Danitrol; Fenpropanate; Herald; Meothrin; Rody; S 3206; SD 41706; WL 41706; XE-938; Fenpropathrine; Kilumal; and Ortho Danitol (Kegley et al. 2012). In this chapter, we will discuss fenpropathrin's uses, its mechanism of toxic action, physical and chemical properties, environmental fate, and ecotoxicology.

1.1 Uses in the United States

Fenpropathrin was first used as an acaricide for both agricultural and horticultural applications. In addition to combating mites, it is a general insecticide that targets representatives of the orders Lepidoptera, Coleoptera, Hemiptera, Homoptera,

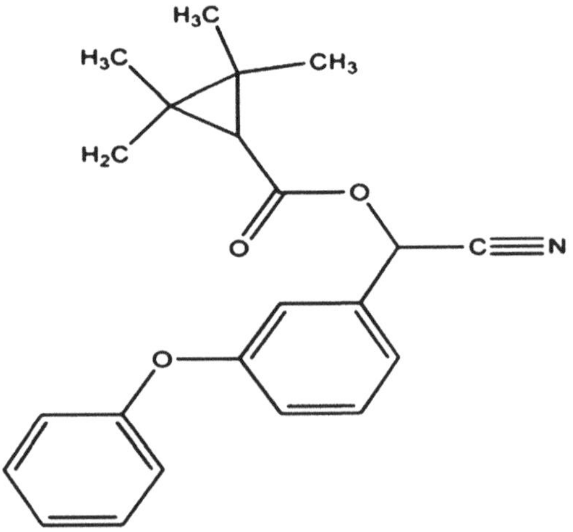

Fig. 1 Structure of fenpropathrin

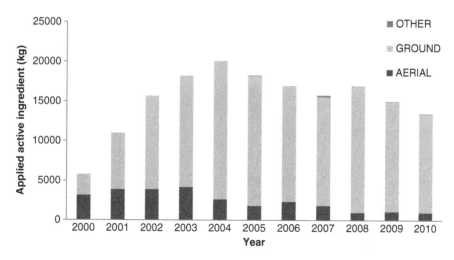

Fig. 2 Total weight of fenpropathrin used (a.i. in kg) in California by application type, 2000–2010 (adapted from CDPR 2012a)

Thysanoptera, Diptera, and Orthoptera (HSDB 2012). In combination with malathion, it has been used to control whitefly (CDPR 2012a). Additionally, this pyrethroid has been especially effective against lygus bug outbreaks on strawberries, although an increase in pyrethroid tolerance among pest populations has led to a decrease in its use (Zalmon et al. 2008).

Although no national use data on fenpropathrin is available, the U.S. Department of Agricultures (USDA) Agricultural Marketing Service (AMS) compiles pesticide residue results from State and Federal agencies through their Pesticide Data Program (PDP). From 2000 to 2010, 71,964 samples of fruit and vegetables were collected and analyzed for the agent (USDA 2010); some 860 (1.2%) tested positive for residues (USDA 2010). The year with the highest frequency of detection was 2004 with 183 (2.3%) samples testing positive for residues out of 7,930 samples analyzed. Additionally, 1,914 treated and 1,908 untreated water samples were analyzed with no detections reported (USDA 2010).

In California, while fenpropathrin is used on a variety of edible and ornamental crops, it is most widely applied to cherries, grapes (both wine and table), oranges, cotton, tomatoes, and strawberries (CDPR 2012a). In 2010, there were 4,169 agricultural applications in California with a total reported weight of 13,434 kg (CDPR 2012a); 153 (3.66%) were applied using aerial spraying techniques for a total weight of 964 kg (7.1%), with the remainder being applied via ground equipment (Fig. 2). The seven crops mentioned above accounted for 85% of the total weight of fenpropathrin applied in California during 2010 (CDPR 2012a), and 83% of that applied from 2006 to 2010 (Fig. 3). The counties with the highest reported use were Tulare, San Joaquin, Kern, Fresno, and Monterey; they were responsible for 66% of the insecticide applied in California in 2010 (CDPR 2012a). In addition, these five counties accounted for 64% of the agent applied on average from 2006 to 2010

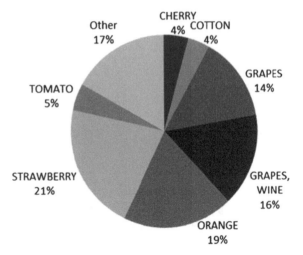

Fig. 3 Five year average of fenpropathrin use on seven crops in California, 2006–2010 (adapted from CDPR 2012a)

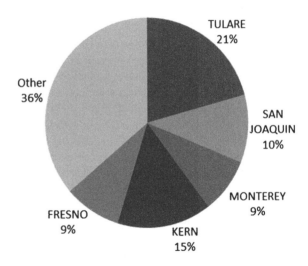

Fig. 4 Reported average weight of fenpropathrin applied in the five counties with the highest use in California, 2006–2010 (adapted from CDPR 2012a)

(Fig. 4). Over the past decade pyrethrins and the pyrethroids have seen an increase in use as a replacement to organophosphorus insecticides because of their reduced toxicity to mammals and birds (Bailey 2009). However, the statewide use of this pyrethroid has followed a decreasing trend since 2004 (Fig. 2).

1.2 Mode of Action

Fenpropathrin is classified as a type II pyrethroid, distinguished from naturally occurring pyrethrins and type I pyrethroids by structural differences resulting from the addition of an α-cyano group. Both pyrethroid types share a common mechanism of prolonging the open time of voltage-gated sodium channels on nerve axons, thus altering membrane excitability (Bailey 2009). This can cause depolarization and the creation of multiple action potentials, leading to choreoathetosis with salivation or CS-Syndrome. The addition of the α-cyano group increases fenpropathrin's insecticidal activity by increasing its affinity for sodium channels (Bailey 2009). Furthermore, the presence of the α-cyano group may broaden the range of voltage-gated channels affected by type II pyrethroids, although it is unclear whether the effects on other channels contribute to the toxic action of this pyrethroid.

2 Physicochemical Properties

Pyrethroids are synthetic esters of naturally occurring pyrethrins. Fenpropathrin (Fig. 1) exists as a racemic mixture that is a yellow-brown solid at standard conditions. It has a density higher than water and low water solubility (Table 1). Similar to other pyrethroids, it is quite hydrophobic and sorbs strongly to particulates and organic matter in aqueous conditions. A high organic carbon partition coefficient (K_{oc}) suggests limited mobility within soil, and fenpropathrin's vapor pressure and

Table 1 Physico-chemical properties of fenpropathrin

Pure physical state	Yellow to brown solid[a]
CAS number	64257-84-7 (racemate)[a]
	39515-41-8 (unstated stereochemistry)[a]
DPR chemical code	2234[b]
Molecular weight (g/mol)	349.4[a]
Molecular formula	$C_{22}H_{23}NO_3$[a]
Density (g/mL)	1.15 (at 25°C)[a]
Melting point (°C)	47.5[a]
Water solubility (mg/L)	0.33 (at 20°C)[a]
Vapor pressure (mPa)	0.76 (at 25°C)[a]
Octanol-water partition coefficient (log K_{ow})	6.00 (at 20°C)[a]
Henry's law constant at 25°C (Pa m^3 mol^{-1})	1.82
Bioconcentration factor (BCF)	1,100[c]
Soil adsorption coefficient (K_{oc})	>4,000[c]
Soluble in organic solvents (mg/L at 20°C)	
Acetone	>500[a]
Acetonitrile	>500[a]
Cyclohexanone	>500[a]
Ethyl acetate	>500[a]
Methanol	>500[a]
Xylene	>500[a]

[a]PPDB
[b]CDPR (2012b)
[c]Laskowski (2002)

Henry's law constant indicate a moderate volatility. A high octanol–water partition coefficient (K_{ow}) and bioconcentration factor suggests a strong tendency for the pesticide to bioaccumulate (Table 1).

3 Environmental Fate

3.1 Air

Fenpropathrin has a vapor pressure of 0.76 mPa, and when released into the atmosphere is predicted to exist in both vapor and particulate phases. Vapor-phase fenpropathrin is degraded via reaction with photochemically produced hydroxyl radicals; the half-life ($t_{1/2}$) is estimated to be 22 h (Table 2; HSDB 2012). Additionally, particulate-phase fenpropathrin will be removed from the atmosphere through deposition, both wet and dry, or degraded through direct photolysis and is thus predicted not to persist for more than a few days (HSDB 2012). However, at an experimental farm in Colmar, France, atmospheric concentrations of the pyrethroid were observed to persist long after application (Millet et al. 1997). Between 1991 and 1993, 18 fog water, 31 rainwater, and 17 gaseous and particulate atmospheric samples were analyzed during both the application season and throughout the year;

Table 2 Degradation half-life values for fenpropathrin in various media

Reference	Media	$t_{1/2}$ (d)
HSDB (2012)	Air (estimated)	0.91
Sakata et al. (1992)	Light clay soil	24.5
Sakata et al. (1992)	Sandy clay loam soil	18
Takahashi et al. (1985a)	River water	19
Takahashi et al. (1985a)	Sea water	11
Takahashi et al. (1985a)	Humic acid solution	42
Takahashi et al. (1985a)	2% Aqueous acetone	1
Takahashi et al. (1985a)	Distilled water	>42
Takahashi et al. (1985a)	Light clay soil	1
Takahashi et al. (1985a)	Sandy loam soil	4
Takahashi et al. (1985a)	Sandy clay loam soil	5
Takahashi et al. (1985a)	Surface of Mandarin orange leaf	4
Mikami et al. (1985)	Cabbage	11.5
Akhtar et al. (2004)	Soil with fertilizer	>56
Dureja (1990)	Sandy loam soil	3.5
Takahashi et al. (1985b)	Buffer solution pH 9.4 at 25°C	<3
Takahashi et al. (1985b)	Buffer solution pH 8.9 at 40°C	<3
Takahashi et al. (1985b)	Buffer solution pH 8.0 at 55°C	<3
Takahashi et al. (1985b)	Buffer solution pH 6.0 at 55°C	37.6
Takahashi et al. (1985b)	Buffer solution pH 1.9 at 40°C	1,280
Al-Makkawy and Madbouly (1999)	Tap water	2.5

concentrations across all media ranged from non-detectable to 900 μg/L. The highest mean value was detected in particle-bound fog water samples (70 μg/L), while the lowest was detected in samples of rainwater (0.26 μg/L). With the exception of particulate-phase air samples, all maximum concentrations were detected between November and January—when pesticides are not applied. The highest concentration of fenpropathrin in particulate-phase air was collected in July of 1992. Regardless of the season, the pyrethroid was detected in one or more atmospheric phases. Although both use and climatic conditions may alter the degradation profile, this study suggests fenpropathrin residues may exist in the atmosphere for extended periods of time after application.

3.2 Soil

Fenpropathrin is nonpolar and therefore has low water solubility. This property causes it to sorb strongly to soils and organic matter to avoid contact with water for thermodynamic reasons. Consequently, the insecticide should have a low potential for leaching through soils to groundwater. Its Henry's law constant indicates that volatilization from wet soil is possible but is unlikely due to its strong sorption capacity. Additionally, volatilization from dry soil is limited by its moderate vapor pressure (HSDB 2012).

Fenpropathrin's potential for leaching has been examined in light clay, sandy clay loam, clay loam, and sand by Sakata et al. (1990). Soils were treated and labeled with ^{14}C and used either immediately or aged for 4 weeks under dark, aerobic conditions. Under both treatments <1.0% of the pyrethroid was detected in the eluates from the light clay, sandy clay loam, and clay loam soils. However, 21–47% of the applied ^{14}C was detected in the eluate of the sandy soil. While fenpropathrin has a high soil sorption coefficient, this study highlights the influence of soil type on its potential for leaching. Its low sorption to sand is likely due to a low organic matter content (<0.1%); being nonpolar, fenpropathrin will sorb best to soils of high organic matter content. Thus, in the absence of organic matter, the pesticide is slightly more likely to percolate into groundwater. In California, 88 sediment samples analyzed had no detectable concentration of fenpropathrin (CDPR 2012c).

3.3 Water

Possessing both low water solubility and a high K_{oc} value, fenpropathrin sorbs strongly to suspended particulates and organic matter in the water column. Being soluble in a wide range of organic solvents, it also tends to bioaccumulate in living organisms. Additionally, with moderate vapor pressure and Henry's law constant values and low water solubility, this pyrethroid may volatilize from water surfaces to a limited extent.

Few studies describing the activity of this agent in the water column are available because of fenpropathrin's sorption tendency. Potential risks to aquatic invertebrates and fish exist because of the high toxicity of fenpropathrin observed in laboratory studies (Solomon et al. 2001). However, under field conditions, the bioavailability of the insecticide is low and its degradation into nontoxic products is relatively rapid (see below), limiting its exposure to living organisms. It is predicted to adhere to soil and particulate matter, degrade at a moderate to rapid rate, and have little solubility in aqueous systems. In California, 432 samples taken from surface water around the state did not contain fenpropathrin at detectable levels (CDPR 2012c).

4   Biotic Degradation

Fenpropathrin's degradation in soil appears to be largely mediated by aerobic microbes. Chapman et al. (1981) tested its persistence in sterile and organic soils and sterile and natural mineral soils. All soils were spiked with 1 mg/kg of the agent and aged under aerobic conditions for 8 weeks. Results indicated degradation was more rapid in natural organic and mineral soils (8% and 2% of the parent remaining, respectively), compared to sterilized organic and mineral soils (83% and 94% of the parent remaining, respectively).

Sakata et al. (1990) conducted experiments on the degradation of this agent in light clay and sandy clay loam soils under aerobic, anaerobic, and sterile conditions. Soils were pre-incubated for 2 weeks in the dark before being treated with the insecticide (at 1 ppm) under aerobic conditions. The $t_{1/2}$ of the insecticide was estimated at 11 and 17 days for the light clay and sandy clay loam soils, respectively. Under anaerobic conditions, soils were again treated at 1 ppm and incubated for 2 weeks under nitrogen gas; after 8 weeks residues were stable, and fenpropathrin was detected at concentrations between 0.85 and 0.87 ppm. To achieve sterile conditions soils were autoclaved and incubated for 8 weeks. At the end of the study residue concentrations were found to be stable at 0.93 ppm for both soils. Sakata et al. (1992) also performed a degradation study in two Japanese upland soils under aerobic conditions. Both a light clay soil and a sandy clay loam were incubated for 2 weeks in the dark and then were brought to a concentration of 1 ppm. Soils were kept in the dark at constant soil moisture during the study period. Fenpropathrin had an estimated $t_{1/2}$ of 24.5 days in the light clay soil and 18 days in the sandy clay loam (Table 2).

Akhtar et al. (2004) examined the persistence of this pyrethroid in soil when used in combination with fertilizers, and the resultant effects on microbial growth. Soils were obtained from vegetable fields and spiked with 100 or 1,000 ppm of fenpropathrin. The fertilizer, consisting of diammonium phosphate, salt of potash, urea, and Polydol™, was applied at 1%. Microbial cultures made from 1 g of un-autoclaved soil and 100 mL of sterilized water were added as a treatment group to identify changes in degradation from microbial processing. Throughout 2 months of study, no degradation was observed in any treatment. Changes in soil pH had no effect on degradation, and at the 100 ppm concentration the pyrethroid had no adverse effects on microbial growth. However, at 1,000 ppm changes in abundance of some colonies were observed. With such high pesticide concentrations, microbial degradation capacity may have been inhibited. Furthermore, no treatment group containing only soil and culture was analyzed for microbial community composition, and therefore, soil microbial community structure with and without the addition of fenpropathrin cannot be compared.

The fate of fenpropathrin has also been studied in various plants (Mikami et al. 1985). For instance, the metabolic fate in cabbage was studied by radiolabeling the α-cyano, benzyl, and cyclopropyl rings and applying foliar treatments in a greenhouse for 42 days. The agent rapidly penetrated cabbage plants and was metabolized with a $t_{1/2}$ of 11–12 days (Table 2); ester hydrolysis was thought to be the major pathway. Subsequent metabolism involved cyano hydrolysis to the respective amide and carboxylic acids, hydroxylation of one or both gem-dimethyl groups, leading to oxidation into a carboxylic acid, and hydroxylation of the phenoxy group. Hydrogen cyanide was released upon hydrolysis of the ester linkage and the remainder rapidly converted into amino acids and dipeptides. Less than 1.2% of the pyrethroid was recovered outside of the treated portions of the plant, indicating that the pesticide does not readily translocate. Additionally, bean plants grown to maturity in soils previously treated with fenpropathrin exhibited no significant uptake or retention (Mikami et al. 1985).

5 Abiotic Degradation

5.1 Photolysis

The rate of photolytic degradation of fenpropathrin on soil is dictated by light intensity, soil moisture, soil acidity, and composition. Reaction kinetics on soil surfaces and the resulting degradation products have been studied by multiple investigators (Dureja 1990; Katagi 1993; Takahashi et al. 1985a); photolysis products include α-carbamoyl-3-phenoxybenzyl-2,2,3,3-tetramethylcyclopropanecarb oxylate and 3-phenoxybenzoic acid (Fig. 5). Dureja (1990) examined photolysis on sandy loam soil under both tropical sunlight and UV light conditions. The insecticide was applied at a rate of 1.3 μg/cm^2 to soil samples, which were then exposed to sunlight for 15 days; the estimated $t_{1/2}$ was calculated to be 3–4 days (Table 2). Additionally, Takahashi et al. (1985a) applied this pyrethroid to three types of soil: light clay, sandy loam, and sandy clay loam. Treated soils were set on thin layer plates (TLP), which were subsequently exposed to natural sunlight for 14 days. At the end of the study period, the calculated half-lives for this agent were 1, 4, and 5 days, respectively (Table 2).

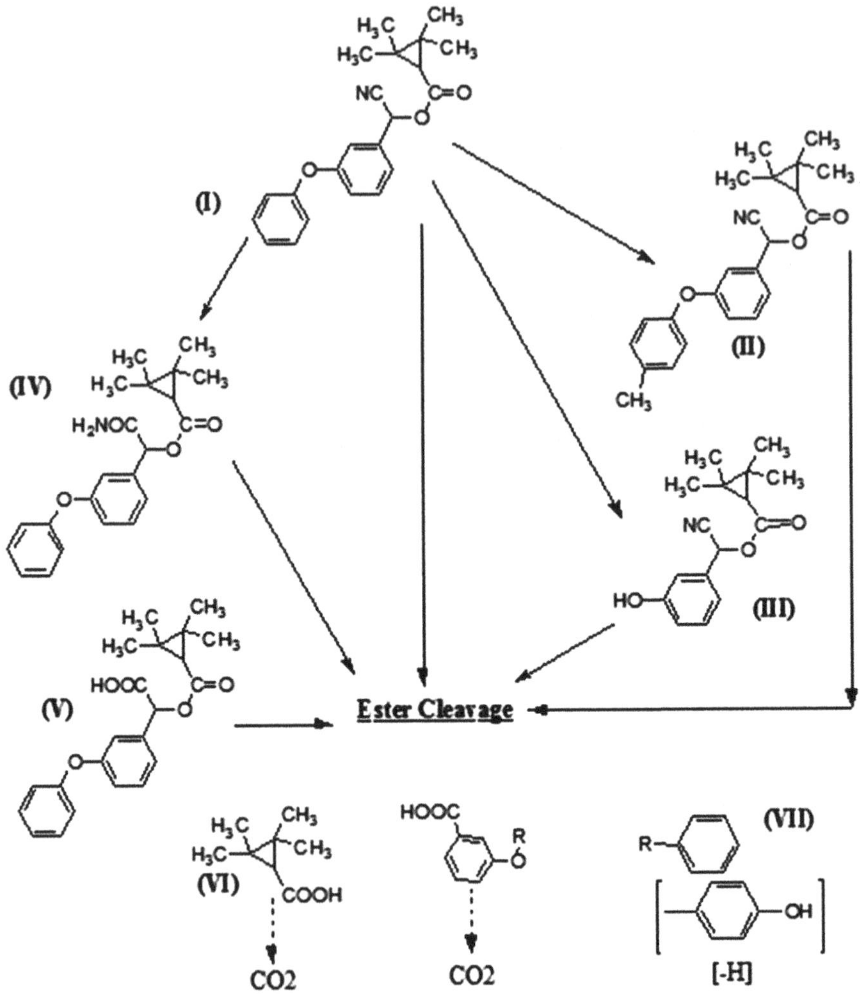

Fig. 5 Degradation pathways of fenpropathrin in soil: II, (RS)-α-cyano-3-hydroxybenzyl 2,2,3,3 tetramethylcyclopropanecarboxylate; III, (RS)-α-cyano-3-(4-hydroxyphenoxy) benzyl 2,2,3,3 tetramethylcyclopropanecarboxylate; IV, (RS)-α-carbamoyl-3-phenoxybenzyl 2,2,3,3 tetramethylcyclopropanecarboxylate; V, (RS)-α-carboxy-3-phenoxybenzyl 2,2,3,3 tetramethylcyclopropanecarboxylate; VI, 2,2,3,3 tetramethylcycloporpanecarboxylic acid; and VII, 3-phenoxybenzoic acid (adapted from Sakata et al. 1990)

Soil moisture content also affects photolytic rate. Katagi (1993) analyzed the effects of soil moisture content and UV irradiation on the degradation of fenpropathrin. Clay loam and two loam soil TLPs were prepared, and soil moisture was adjusted to levels ranging from 0% (oven dried) to 100% (saturation). Samples were fortified with the agent and continuously irradiated with artificial light for 14 days. The degradation profile changed significantly with soil moisture content—when it exceeded 30%, an almost constant degradation rate of $0.7–1.0 \times 10^{-2}$ day^{-1} was observed. However, degradation rates in soils with moisture content below 17% increased significantly to 0.92 day^{-1}. In the oven-dried soil, 90% of fenpropathrin degraded within the first 3 days, with a $t_{1/2}$ less than a day. Enhanced UV irradiation showed little effect on degradation rates. Hydration of the α-cyano group followed by hydrolysis of the amide group was predominant in soils with moisture content under 17%.

The surface acidity of clay is known to increase as moisture content decreases via polarization of water molecules by exchangeable cations (Theng 1982; Voudrias and Reinhard 1986). The predominant reactions in soils having high moisture content (50%) were cleavage of the ester linkage and hydroxylation. Degradation pathways and rates were also influenced by soil characteristics such as clay quantity and species, and organic matter; primary break-down products were (RS)-α-carbamoyl-3-phenoxybenzyl 2,2,3,3 tetramethylcyclopropanecarboxylate, (RS)-α-carboxy-3-phenoxybenzyl 2,2,3,3 tetramethylcyclopropanecarboxylate, 3-phenoxybenzoic acid, and CO_2. Katagi concluded that degradation profiles were mainly dependent on the acidity of the clay as a function of soil moisture, clay species, and organic matter and not photolysis. The relationship of these parameters created an environment in which the insecticide was readily degraded because of the accessibility to its α-cyano group.

5.2 Hydrolysis

Fenpropathrin is fairly stable in aquatic systems under acidic conditions, with half-lives ranging from 38 to 1,280 days (Takahashi et al. 1985b). Under basic conditions (pH 9.4 at 25°C) it dissipates more quickly, with observed half-lives of less than 3 days (Table 2). Takahashi et al. (1985b) also studied its hydrolysis in both river and sea water at varying temperatures and pH. Buffer solutions used ranged from pH 1.9 to 10.4 and temperature from 25 to 65°C. Fenpropathrin was found to be unstable under basic conditions, with an estimated $t_{1/2}$ of less than 3 days at pH 9.4 and above 25°C. The rate of hydrolysis and the formation of degradation products increased as buffer solutions increased in pH. Additionally, by keeping the pH constant and increasing the temperature, the rate of hydrolysis was increased. However, under acidic conditions it was found to be relatively stable, with a half-life ranging from 37.6 days at pH 6 and 55°C to 1,280 days at pH 1.9 and 40°C. Takahashi found that regardless of pH and temperature, the formation of 2,2,3,3-tetramethylcyclopropanecarboxylic acid (Fig. 5) was more rapid than the formation of (RS)-α-carbamoyl-3-phenoxybenzyl 2,2,3,3-tetramethylcyclopropanecarboxylate (Fig. 5). Cleavage of the ester bond via hydrolysis was more rapid with neutral and base-catalyzed reactions, when compared to hydrolysis of the α-cyano group, regardless of pH and temperature (Takahashi et al. 1985b). Under neutral and acidic conditions, fenpropathrin remains stable.

Takahashi et al. (1985b) exposed fenpropathrin to natural sunlight in both river and sea water samples for 2 weeks and found it to have half-lives of 2.7 and 1.6 weeks, respectively. The accelerated photodegradation observed in both river and sea water, when compared to distilled water (>6 weeks; Table 2), may be attributed to natural photosensitizers such as humic acids, tryptophan, and tyrosine present in natural waters (Takahashi et al. 1985b). Additionally, Takahashi et al. (1985b) found that breakdown products in natural waters consisted primarily of 2,2,3,3-tetramethylcycloporpanecarboxylic acid, (RS)-α-carboxy-3-phenoxybenzyl 2,2,3,3-tetramethylcyclopropanecarboxylate, and 3-phenoxybenzoic acid, suggesting that the ester bond of fenpropathrin is more susceptible to hydrolysis than the α-cyano group (Fig. 5).

6 Ecotoxicology

6.1 Insects

Fenpropathrin is 1,000–2,250 times more toxic to insects than to mammals (Bailey 2009; Cage et al. 1998), and there are a number of factors that contributes to the difference in toxicity. For instance, Song and Narahashi (1996) found that the sensitivity of voltage-gated sodium channels to pyrethroids is much greater in insects than in mammals. Variation in sodium channel sub-units and differences between regions of the neuronal membrane may explain the differences, although the interaction between pyrethroids and sodium channels is not fully understood (Bailey 2009). Additionally, neuronal poisoning may occur more rapidly in insects because of their small relative size and decreased time available for detoxification (Song and Narahashi 1996).

Fenpropathrin's toxicity to insects has been studied by a number of investigators (Bellows and Morse 1993; Michaud 2002; Michaud and Grant 2003). Michaud and Grant (2003) tested its toxicity to nontarget insects representing four orders: Coleoptera, Neuroptera, Hymenoptera, and Hemiptera. Exposures were conducted at the standard field rate for citrus in Florida of 309 ppm, a 10-fold, and a 100-fold dilution. Application was by topical spray during the first instar of development or through exposing the second instar to foliar residue for 24 h. Complete mortality was observed for four species representing Coleoptera and a single species from Neuroptera exposed during the first instar at the field rate concentration. The four Coleopterids experienced 100% mortality at the 10-fold dilution and mortality ranged from 0 to 75% at the 100-fold dilution. Single species from the Orders Hymenoptera and Hemiptera were treated during the second instar; at field rate concentrations mortality was >96%. Mortality was 64% and 12% at the 10-fold dilution and 25% and 0% at the 100-fold dilution for the Hemipterid and Hymenopterid species, respectively.

The pyrethroids are highly toxic to both target and nontarget insects. Bellows and Morse (1993) found that the pyrethroids were the most toxic to two Coleopterid

Table 3 Ecotoxicology data for selected species

Common name	Species	Exposure time (h)	LC$_{50}$ (µg/L)
Waterflea[a]	*Daphnia magna*	48	0.5
Honey Bee[b]	*Apis mellifera*	48	0.05
Bluegill[a]	*Lepomis macrochirus*	96	2.3
Rainbow Trout[a]	*Oncorhynchus mykiss*	96	2.3
Channel Catfish[a]	*Ictalurus punctatus*	96	5.5
Sheepshead Minnow[c]	*Cyprinodon variegatus*	96	3.1
Grass Carp[c]	*Ctenopharyngodon idella*	48	3.59
Western Mosquitofish[c]	*Gambusia affinis*	48	1.3
Mallard Duck[c]	*Rana limnocharis*	192	9.026×10^6
Opossum Shrimp[c]	*Americamysis bahia*	96	0.021
Fiddler Crab[c]	*Uca pugilator*	96	5.2

[a]Kegley et al. (2012)
[b]Valent (2008)
[c]U.S. EPA (2007)

species, followed by the carbamates and organophosphates, with fenpropathrin being the most toxic of the pyrethroids tested. As a result of pyrethroid exposure, developmental times of insects increased, negatively affecting fitness. However, Michaud (2002) noted that larvae of *H. axyridis* showed decreased developmental times when exposed to leaf residues of fenpropathrin, resulting in incomplete development. Many nontarget species play a vital role in predation, and thus on the control of crop pests. Additionally, because fenpropathrin is a highly toxic contact poison, it may negatively impact honey bee (*Apis mellifera)* and other insect populations and inhibit their pollination services for agriculture; the acute contact LD$_{50}$ for honey bees is 0.05 µg/bee (Table 3). This falls in the middle of the range of pyrethroid LD$_{50}$ values, compared to 0.029 µg/bee for contact application of permethrin and 0.103 and 0.044 µg/bee for cypermethrin and bifenthrin 48 h after ingestion, respectively. Additionally, the high toxicity and fat solubility of the pyrethroids relative to the neonicotinoids may present a greater risk to the honey bee. Further research on how to limit and mitigate the effects of pyrethroid exposure on nontarget insect populations is needed.

6.2 Aquatic Organisms

Fenpropathrin's mode of action and high bioconcentration factor (1,100) make it potentially highly toxic to fish and other aquatic organisms; exposure may arise from urban, agricultural, or mixed-use sources. It is more toxic at cooler temperatures, and thus more toxic to cold water than warm water fish, but the toxicity of pyrethroids is little affected by pH or water hardness (Cage et al. 1998; Mauck et al. 1976). The 96-h LC$_{50}$ for rainbow trout and channel catfish are 2.3–5.5 µg/L, respectively. Toxicity values for additional species are listed in Table 3.

The potential for bioaccumulation of fenpropathrin has been examined using the fish *Tilapia nilotica* and water from the Nile River (Al-Makkawy and Madbouly 1999). Upon application to water at a rate of 1 μg/L, fenpropathrin degraded gradually and was undetectable after 28 days; the predicted $t_{1/2}$ in water was 2.5 days (Table 2). Maximum residue concentrations in the heads and flesh of *T. nilotica* were 130 ppb and 7 ppb at 3 days, respectively.

Acute toxicity of the agent was also studied in the grass carp, *Ctenopharyngodon idellus*. Wu et al. (1999) employed laboratory fresh water systems, with the addition of dissolved humic material (DHM), to determine its influence on the bioavailability and toxicity of the insecticide. Increasing concentrations of DHM led to a decrease of free aqueous residues. Additionally, carp mortality was inversely correlated with humic material content, as would be expected; 8-h LC_{50} values were 2.51, 3.72, and 4.68 μg/L in 0, 1, and 10 mg/L DHM, respectively. The inverse relationship is probably due to sorption of the pyrethroid by the humus, as it reduces free pesticide available to diffuse through biological membranes. Similar results have also been observed for *Daphnia magna*, with a 20–80% reduction in mortality when pyrethroid exposure occurred in the presence of humic material (Day 1991).

Ding et al. (2011) evaluated the presence and toxicity of fenpropathrin in the Central Valley of California. Three different sediment types were spiked with the insecticide for 10-day bioassays involving *Hyalella azteca* (third instar); a single sediment type was used for similar tests involving *Chironomus dilutus* (juvenile). Sediments were spiked with the insecticide, covered with aluminum foil and aged for 12 days at 4°C before use. The insecticide persisted in all sediments, with degradation rates ranging from 1 to 12% over the 10-day test. The estimated 10-day LC_{50} for *C. dilutus* was 177 ng/g, while it ranged between 11 and 40 ng/g for *H. azteca*. In California, fenpropathrin residues have been observed at concentrations up to 19.4 ng/g in sediment samples, indicating the prospect photodegradation products (RS)-α-carbamoyl-3-phenoxybenzyl 2,2,3,3-tetramethylcyclopropanecarboxylate, 2,2,3,3 tetramethylcycloporpanecarboxylic acid, and 3-phenoxybenzoic acid are markedly less toxic than the parent, with killifish 48-h LC_{50} values of more than 10 mg/L (Takahashi et al. 1985a).

6.3   Birds

Similar to other pyrethroids, fenpropathrin displays low toxicity to terrestrial game birds. Dietary LC_{50} values during 192-h exposures for mallard ducks and northern bobwhite quail have been reported as 9,026 and >10,000 mg/kg, respectively (Tomlin 1994). Although there is the potential for the insecticide to bioaccumulate, it is thought that the insecticide rapidly transforms into nontoxic metabolites for efficient excretion.

6.4 Mammals

Fenpropathrin is relatively nontoxic to mammals due to the time it takes to reach the central nervous system; the LD_{50} for a male rat is 70.6 mg/kg in corn oil (Tomlin 1994). In general, there is little concern for acute toxicity to mammals from pyrethroids, because these pesticides rapidly biotransform through ester hydrolysis and oxidation reactions. These processes result in the formation of inactive 3-phenoxybenzoic acid and additional products (McEvoy 2000). Although the major inactivation pathway of pyrethroids in mammals is through metabolic processing, the inverse relationship between temperature and pyrethroid toxicity supports their being low concern for fenpropathrin toxicity at mammalian body temperatures.

7 Summary

Fenpropathrin was the first of the light-stable pyrethroids to be synthesized. It is a broad-spectrum insecticide that is used on numerous crops throughout the USA. It is relatively nonpolar, contributing to its low water solubility, moderate volatility, high octanol–water and organic carbon partition coefficients, and significant sorption to both particulate and organic matter. It degrades in the environment primarily through photolysis, hydrolysis, and aerobic metabolism in both water and soil. Volatilization has been demonstrated, and residues have been detected in air samples for weeks after application. It has low mobility in soil and consequently a low potential for leaching to groundwater. The rate of photolysis in soil was determined to be a function of soil moisture and soil acidity, which decreases the half-life when soil moisture and acidity are increased. Under these conditions, half-lives can range from 1 to 5 days.

Fenpropathrin in soil is also degraded via aerobic microbes, and half-lives under aerobic conditions were found to be between 11 and 17 days. It remains stable under anaerobic and sterile conditions, as no significant degradation was measured after 8 weeks. Photodecomposition of the insecticide in river and sea water is moderate with $t_{1/2}$ estimates of 2.7 and 1.6 weeks, respectively. Hydrolysis proceeds primarily through neutral- and base-catalyzed reactions, with observed half-lives in natural waters ranging from less than 3 days to 2.7 weeks. Under acidic conditions, fenpropathrin was shown to remain relatively stable with half-lives estimated to be as long as 8,520 days; degradation occurs primarily through ester bond cleavage.

Fenpropathrin's properties influence its high degree of toxicity to fish and aquatic invertebrates as well as to nontarget terrestrial insects. This toxicity and the ability to bioaccumulate in organisms are mitigated in the field by fenpropathrin's efficient metabolic degradation into nontoxic products, followed by rapid excretion. In addition, organic material in water may interact with fenpropathrin, reducing its bioavailability. Consequently, under field conditions there is relatively low concern that fenpropathrin will cause toxicity to mammals, birds, and fish. However, fenpropathrin's high aquatic invertebrate toxicity and potential to persist in anaerobic environments suggest measures should be taken to prevent excessive runoff into surrounding waterways.

Acknowledgments Support was provided by the Environmental Monitoring Branch of the California Department of Pesticide Regulation (CDPR), California Environmental Protection Agency. The statements and conclusions are those of the authors and not necessarily those of CDPR. The mention of commercial products, their source, or their use in connection with materials reported herein is not to be construed as actual or implied endorsement of such products.

References

Akhtar S, Gilani S, Hasan N (2004) Persistence of chlorpyrifos and fenpropathrin alone and in combination with fertilizers in soil and their effects on soil microbes. Pak J Bot 36:863–870

Al-Makkawy H, Madbouly M (1999) Persistence and accumulation of some organic insecticides in Nile water and fish. Resour Conserv Recy 27:105–115

Bailey JE (2009) U.S. Environmental Protection Agency, FIFRA Scientific Advisory Panel. Meeting minutes. http://www.epa.gov/scipoly/sap/meetings/2009/june/061609minutes.pdf. Accessed 22 Apr 2012

Bellows TS Jr, Morse JG (1993) Toxicity of pesticides used in citrus to *Aphytis melinus* DeBach (Hymenoptera: Aphelinidae) and *Rhizobius lophanthae* (Blaisd.) (Coleoptera: Coccinellidae). Can Entomol 125:987–994

Cage SA, Bradberry SM, Meachem S, Vale JA (1998) National Poisons Information Service. http://www.intox.org/databank/documents/chemical/fenprop/ukpid64.htm. Accessed 7 Feb 2012

CDPR (2012a) Pesticide use reporting. Database. http://www.cdpr.ca.gov/docs/pur/purmain.htm. Accessed 27 March 2012

CDPR (2012b) Product label. Database. http://www.cdpr.ca.gov/docs/label/labelque.htm. Accessed 27 March 2012

CDPR (2012c) Surface water monitoring. Database. http://www.cdpr.ca.gov/docs/emon/surfwtr/surfcont.htm. Accessed 27 Aug 2012

Chapman R, Tu C, Harris C, Cole C (1981) Persistence of 5 pyrethroid insecticides in sterile and natural, mineral and organic soil. Bull Environ Contam Toxicol 26:513–519

Davies JH (1985) The pyrethroids: a historical introduction. In: Leahey JP (ed) The pyrethroid insecticides. Taylor and Francis, London, pp 1–41

Day KE (1991) Effects of dissolved organic carbon on accumulation and acute toxicity offenvalerate, deltamethrin and cyhalothrin to *Daphnia magna* (Straus). EnvironToxicol Chem 10:91–101

Ding Y, Weston D, You J, Rothert A, Lydy M (2011) Toxicity of sediment-associated pesticides to *Chironomus dilutus* and *Hyalella azteca*. Arch Environ Contam Toxicol 61:83–92

Dureja P (1990) Photodecomposition of pyrethroid insecticide fenpropathrin. Pesticides 1989:31–33

HSDB (2012) Fenpropathrin hazardous substances databank. http://toxnet.nlm.nih.gov/cgi-bin/sis/htmlgen?HSDB. Accessed 12 Feb 2012

Katagi T (1993) Effect of moisture content and UV irradiation on degradation of fenpropathrin on soil surfaces. J Pest Sci 18:333–341

Kegley SE, Hill BR, Orme S, Choi AH (2012) Pesticide action network pesticide database http://www.pesticideinfo.org. Accessed 25 Feb 2012

Laskowski DA (2002) Physical and chemical properties of pyrethroids. Rev Environ Contam Toxicol 174:49–170

Mauck W, Mayer F, Holz D (1976) Simazine residue dynamics in small ponds. Bull Environ Contam Toxicol 16:1–8

McEvoy GK (ed) (2000) American Hospital Formulary Service—Drug information. American Society of Health-System Pharmacists, Bethesda, MD, p 3203

Michaud JP (2002) Relative toxicity of six insecticides to *Cycloneda sanguinea* and *Harmonia axyridis* (Coleoptera: Coccinellidae). J Entomol Sci 37:82–93

Michaud JP, Grant AK (2003) IPM-compatibility of foliar insecticides for citrus: indices derived from toxicity to beneficial insects from four orders. J Insect Sci 3:18–29

Mikami N, Baba Y, Katagi T, Miyamoto J (1985) Metabolism of the synthetic pyrethroid fenpropathrin in plants. J Agric Food Chem 33:980–987

Millet M, Wortham H, Sansui A, Mirabel P (1997) Atmospheric contamination by pesticides: determination in the liquid, gaseous and particulate phases. Environ Sci Pollut Res 4:172–180

PPDB (2009) The pesticide properties database. http://sitem.herts.ac.uk/aeru/footprint/en/index.htm. Accessed 22 March 2012

Sakata S, Yoshimura J, Nambu K, Mikami N, Yamada H (1990) Degradation and leaching behavior of the pyrethroid insecticide fenpropathrin in soils. J Pest Sci 15:363–373

Sakata S, Mikami N, Yamada H (1992) Degradation of pyrethroids optical isomers in soils. J Pest Sci 17:169–180

Solomon K, Giddings J, Maund S (2001) Probabilistic risk assessment of cotton pyrethroids: I. Distributional analyses of laboratory aquatic toxicity data. Environ Toxicol Chem 20:652–659

Song J, Narahashi T (1996) Modulation of sodium channels of rat cerebellar Purkinje neurons by the pyrethroid tetramethrin. J Pharmacol Exp Ther 277:445–453

Takahashi N, Mikami N, Yamada H, Miyamoto J (1985a) Photodegradation of the pyrethroid insecticide fenpropathrin in water, on soil and on plant foliage. J Pest Sci 16:119–131

Takahashi N, Mikami N, Yamada H, Miyamoto J (1985b) Hydrolysis of the pyrethroid insecticide fenpropathrin in aqueous media. J Pest Sci 16:113–118

Theng B (1982) Clay-polymer interactions—Summary and perspectives. Clays Clay Miner 30:1–10

Tomlin CDS (ed) (1994) The pesticide manual—World compendium, 10th edn. The British Crop Protection Council, Surrey, UK, 446

United States Department of Agriculture (2010) Pesticide data program database. www.ams.usda.gov/science/pdp/. Accessed 12 March 2012

United States Environmental Protection Agency (2007) ECOTOX user guide: ECOTOXicology database system. http://www.epa.gov/ecotox/. Accessed 12 March 2012

Valent U.S.A. Corporation (2008) TAME® 2.4 EC material safety data sheet. http://www.valent.com/Data/Labels/0033rev7.pdf. Accessed 2 May 2012

Voudrias E, Reinhard M (1986) Abiotic organic-reactions at mineral surfaces. ACS Symposium Series 323:462–486

Wu W, Xu Y, Schramm K, Kettrup A (1999) Effect of natural dissolved humic material on bioavailability and acute toxicity of fenpropathrin to the grass carp, ctenopharyngodon idellus. Ecotoxicol Environ Saf 42:203–206

Zalmon F, Bolda M, Phillips P (2008) UC IPM pest management guidelines: strawberry. http://www.ipm.ucdavis.edu/PMG/r734300111.html. Accessed 27 May 2012

Use of Land Snails (Pulmonata) for Monitoring Copper Pollution in Terrestrial Ecosystems

Dragos V. Nica, Despina-Maria Bordean, Aurica Breica Borozan,
Iosif Gergen, Marian Bura, and Ionut Banatean-Dunea

Contents

D.V. Nica (✉) • M. Bura • I. Banatean-Dunea
Faculty of Animal Sciences and Biotechnologies, Banat's University of Agricultural Sciences
and Veterinary Medicine, C. Aradului 119, 300645 Timisoara, Romania
e-mail: nicadragos@gmail.com

D.-M. Bordean • A.B. Borozan • I. Gergen
Faculty of Food Products Technology, Banat's University of Agricultural Sciences
and Veterinary Medicine, C. Aradului 119, 300645 Timisoara, Romania

D.M. Whitacre (ed.), *Reviews of Environmental Contamination and Toxicology*,   95
Reviews of Environmental Contamination and Toxicology 225,
DOI 10.1007/978-1-4614-6470-9_4, © Springer Science+Business Media New York 2013

1   Introduction

The term bioindicator defines organisms that respond to a pollutant load with changes in vital functions or accumulate pollutants (Arndt et al. 1987). The use of bioindicator organisms may allow researchers to evaluate the effects of mixtures of pollutants on ecosystems in time and space, depending on the selected species and approach (Hellawell 1986). In contrast, simple instrumental analyses of pollutants can provide extremely precise data about their accumulation in organisms, but they do not clearly reveal how different chemicals interact when they cooccur in complex mixtures (Maynard 2004). Bioindicator organisms are often used in environmental monitoring programs "to assess the condition of the environment, to provide an early warning signal of changes in the environment, or to diagnose the cause of an environmental problem" (Dale and Beyeler 2001). Such studies are designed to mirror the quality of natural environments and can either be passive, when bioindicator organisms are native inhabitants of the ecosystem, or active, when organisms of a known biological past are inserted into the site to be monitored (Markert 2007).

Until recently, plants and other invertebrate groups were most critical to surveillance of contaminated terrestrial ecosystems, and other species such as nematodes, annelids, arthropods, or gastropods played only a complementary role (Lepp and Salmon 1999; Zhang et al. 2007; Strandberg et al. 2009; Wang et al. 2009; Zvereva and Kozlov 2010). However, in the past decade or so, these latter species have attracted enhanced research interest, and therefore, several active (e.g., Gomot de Vaufleury and Pihan 2000; Scheifler et al. 2003; Regoli et al. 2006; Gimbert et al. 2008; Fritsch et al. 2011) and passive biomonitoring-related projects (e.g., Rabitsch 1996; Beeby and Richmond 2002; Notten et al. 2005; Mourier et al. 2011) have been successfully performed with them on both xenobiotic (i.e., Pb, Cd) and physiological heavy metals (i.e., Cu, Zn).

Apart from their ecological role and importance as the most species-rich group of terrestrial mollusks, the land snails' ecological and biological attributes closely adhere to the preconditions of serving as suitable bioindicators. First, these ubiquitous and sinantropic mollusk species are abundant in the wild and have population densities up to 200 individuals per square meter (i.e., *Carychium tridentatum*— Mason 1970). Second, they are easily reared under laboratory conditions (Hodasi 1979; Dan and Bailey 1982; Pawson and Chase 1984; Gray et al. 1985; Gomot and Deray 1987; Jess and Marks 1989; Schuytema et al. 1994; Desbuquois and Madec 1998; Lazaridou-Dimitriadou et al. 1998; Dupont-Nivet et al. 2000; García et al. 2006). Third, their physiological particularities (i.e., dual respiration through lung and tegument, regular ingestion of small amounts of soil and vegetation) allow them to accumulate pollutants through multiple routes of exposure: oral, dermal, and respiratory (Gomot de Vaufleury and Pihan 2000; Regoli et al. 2006; Scheifler et al. 2006; Fritsch et al. 2008).

Several authors (Coughtrey and Martin 1976; Bigliardi et al. 1988/1989; Beeby and Richmond 2003; Viard et al. 2004; Wegwu and Wigwe 2006) have noted that land snails have the ability to tolerate substantive concentrations of the elements they bioaccumulate in their tissues. The physiological basis for this amazing tolerance is the efficient binding of heavy metals ions by specific metallothioneins (Taylor et al. 1988; Dallinger et al. 1989) and their deposition in insoluble intracellular granules (Howard et al. 1981). Snails do, however, suffer biologic effects from metal exposure, such as inhibition of growth and development, as well as alterations to reproductive functions (Russell et al. 1981; Snyman et al. 2000; Gomot de Vaufleury and Kerhoas 2000; Notten et al. 2006). Since land snails are important prey for mammals, birds, and large invertebrates (e.g., Symondson and Lidell 1993; Graveland et al. 1994), they may also be involved in the biomagnification of heavy metals along food chains (Laskowski and Hopkin 1996b). Overall, land snails (*Pulmonata*) are regarded to be a model invertebrate for heavy metal accumulation in terrestrial ecosystems (Gomot de Vaufleury 2000).

Because land snails (*Pulmonata*) are sensitive to pollution and are subject to damage from many pollutants, several pollutants have been candidates for evaluating snails as potential bioindicator organisms. Copper (Cu) was chosen not only because it is an anthropic-induced pollutant that has environmental health consequences but also because people have used it and have been exposed to it on a large scale since the Bronze Age (Hong et al. 1996a, b; Veselý 2000). Even today, large amounts of copper compounds are routinely released to the environment from several key human activities that include industrial and animal production, intensive agriculture, mining operations, siderurgy, and steel manufacturing (Georgopoulos et al. 2001; Warnken et al. 2004; Menezes et al. 2004; Finnie 2006; Hulskotte et al. 2007). As a result, copper release and contamination represents a great matter of concern to much of the World's population (Nogué et al. 2000; Caldentey and Mondschein 2003).

Although there are two types of copper ions, copper (I) and copper (II), the most commonly encountered are copper (II) compounds (Vidal et al. 1999). This may be because copper (I) ions attain stability only under reducing conditions (Neniţescu 1972), something that may be expected within certain cells, but not in an oxidizing environment. At low concentrations, copper ions are trace elements required by all animals as essential constituents of cofactors, enzymes, and proteins (Gadd 1992; Linder and Hazegh-Azam 1996; Stehlik-Tomas et al. 2004). However, the persistence of Cu ions renders them potentially toxic in contaminated ecosystems (Kuffner et al. 2008; Shrivastava 2009). Weser et al. (1979) mentioned at least 30 known copper-based enzymes, all functioning as redox catalysts (e.g., cytochrome oxidase, nitrate reductase) or dioxygen carriers (e.g., hemocyanin); such biologic processes exploit the easy interconversion of copper (I) and copper (II) ions (Lippard and Berg 1994). These same ions, by contrast, are able to produce toxic hydroxyl radicals ($-OH$) in Fenton-like reactions, and as a consequence, they can be toxic to cellular components like proteins, lipids, and DNA (Schumann et al. 2002).

Generally, organisms are exposed to copper via inhalation, food and water consumption, and dermal contact with contaminated air, water, and soil (WHO 2004). Because copper (II) sulfate is a compound that has applications in a wide range of industries (e.g., from pyrotechnics to agriculture), this compound has been used in

Table 1 Critical levels of tolerance to the action of copper that are dependent on the animal species, exposure route and duration, dose, and chemical form of the copper compound involved

Contamination path	Copper dose, form, and effect	Species	Reference
Food	LD_{50} = 472 ppb $CuSO_4$	Rats	Extoxnet (1996)
	LD_{50} = 833 ppb $Cu(OH)_2$	Rats	
	LDLo = 1,000 ppb $CuSO_4$	Pigeons	
	LDLo = 600 ppb $CuSO_4$	Ducks	
	Toxic effect = 11 ppb $CuSO_4$	Humans	
	RDA = 13–19 ppm Cu/day	Humans	WHO (2004)
Dermal	LD_{50} = 5,000 ppb $Cu(OH)_2$	Rats	Extoxnet (1996)
Drinking water	Safe level = 2.0 mg $CuSO_4$/L	Humans	Pimentel (1971)
	NOAEL = 27.20 ppm Cu/day (acute duration oral)		Pizarro et al. (1999)
	MRL = 10 ppm Cu/day (acute duration oral)		ATSDR (2002)
	MCLG = 1.3 mg Cu/L		
Air	PEL (8-h TWA) = 0.1 mg/m^3 Cu (fume)	Humans	ATSDR (2002)
	PEL (8-h TWA) = 1.0 mg/m^3 Cu (dusts and mists)		
	IDLH = 100 mg/m^3 Cu (fumes, dusts, and mists)		
	PM_{10} (standard) = 20 µg/m^3 (annual average)		CEPA (2002)
	$PM_{2.5}$ (standard) = 12 µg/m^3 (annual average)		

Note: LD$_{50}$ median lethal dose, *LDLo* lowest lethal dose, *RDA* average dietary requirements, *NOAEL* acute no-observed-adverse-effect level, *MRL* minimal risk level, *MCLG* maximum contaminant level goal, *PEL* permissible exposure limit, *IDLH* immediately dangerous to life and health, *8-h TWA* average value of exposure over the course of an 8-h work shift, *PM^{10}* particulate matter with diameter less than 10 µm, *PM$_{2.5}$* particulate matter with diameter less than 2.5 µm

many studies as a benchmark of copper's hazard to living organisms and to human health (Brown et al. 1974; Guecheva et al. 2001; Mortazavi and Jafari-Javid 2009). The toxic effects of copper to organisms, under various conditions, are described in Table 1. Short-term exposure of humans to poisoning by Cu ingestion has been associated with vomiting, hypotension, icteric teguments, digestive bleeding, and gastrointestinal distress (Uriu-Adams and Keen 2005; Brewer 2007), whereas long-term exposure induced liver and kidney damage (Singh et al. 2006). Nevertheless, many animals including mammals have efficient mechanisms to regulate copper accumulation, and such mechanisms allow many animal species to counteract excess dietary levels (Bremner 1998; Mercer and Llanos 2003). Several studies have revealed that copper does not bioconcentrate in fish (e.g., Jezierska and Witeska 2006; Karayakar et al. 2010), but has a high potential to do so in mollusks (e.g., Peña and Pocsidio 2008; Lopes et al. 2011). Therefore, these invertebrates were investigated for their potential to serve as bioindicators of copper pollution in both terrestrial and aquatic environments (Oehlmann and Schulte-Oehlmann 2002).

Our purpose in this paper is to address the following practical questions relating to the problems and contradictions of using land snails (*Pulmonata*) as sentinel organisms for monitoring copper contamination:

1. What is the potential for using land snails (*Pulmonata*) as effective bioindicators of copper contamination when they are exposed via contact with air, soil, or food, or via field exposure (i.e., mixed air, soil, and food contamination)?
2. How is copper metabolized by gastropods, and what are the putative pathways for copper accumulation (i.e., in tegument, lungs) and excretion (i.e., urine, mucus)?
3. How is copper accumulation in land snails relevant to their use as indicators of short- and long-term Cu contamination?
4. How have various biomarkers of copper pollution thus far been used in terrestrial ecosystems to perform laboratory and field surveys, what are the limitations associated therewith, and how can such impediments be overcome?

2 Sources of Copper Exposure to Terrestrial Snails (Pulmonata)

Davies and Bennett (1985) reported that Cu is relatively abundant in the Earth's crust and generally exists at a level estimated to be 50 ppm. Copper (II) ions are moderately soluble and disperse into the environment through natural phenomena such as disintegration of minerals and lixiviation—the process of separating soluble from insoluble substances by dissolution in water (Hulmann and Kraft 2003). As a result, copper (II) ions are easily accumulated in soil, water, sediment, and air, as well as in plants and animals.

Because terrestrial gastropods are involved at different trophic levels, both as herbivorous and detritivore organisms (Dallinger et al. 2001a; Scheifler et al. 2003), copper accumulates in their bodies when they consume contaminated vegetation, and when they take up organic debris formed from the decay of organisms that live in polluted ecosystems. In addition, land snails' dual respiration (through lung and tegument) renders them reliable sentinel organisms for monitoring heavy metal emissions (Regoli et al. 2006). Furthermore, contaminated waters originating from either mining activities (Samecka-Cymerman and Kempers 2007) or sewage treatment plants (Deng et al. 2004) enter the snail's bodies via water, food, or soil consumption. As a result, land snails are exposed to the action of heavy metals through multiple routes of contamination.

Gomot de Vaufleury and Pihan (2000) classified routes of exposure to heavy metals as follows: (a) contamination by breathing air; (b) contamination at the soil surface by consuming plants, detritus, soil, and surface water; (c) contamination by contact with soil, water, plant, and detritus. Land snails have different paths of exposure that produce complex mutual actions at the biocenotic level, and therefore in this review, we present research results that are related either to simple copper uptake via air, soil, or food, or to mixed air, soil, and food exposures (field exposure). In Table 1, we provide data on the critical levels of animal and human tolerance to the action of copper; such information may allow a facile assessment of Cu pollution risk in terrestrial ecosystems by using land snails (*Pulmonata*) as bioindicators.

When expressing relative levels of Cu toxicity to different organisms, most researchers have relied on values such as the 50% effective concentration (EC_{50}), the half lethal concentration (LC_{50}) and dose (LD_{50}), the no-observable-effect concentration (NOEC), the lowest-observable-effect concentration (LOEC), average dietary requirements (RDA), lowest lethal dose (LDLo), maximum contaminant level goal (MCLG), or acute no-observed-adverse-effect level (NOAEL) (Table 1). Other terms that are routinely used to express a hazard of Cu or other substances to human health include the permissible exposure limit (PEL), the immediately dangerous to life and health (IDLH) value, the minimal risk level (MRL), the average value for exposure during an 8-h work shift (8-h TWA), and the hazard of particulate matter having diameters <10 μm (PM_{10}) or less than 2.5 μm ($PM_{2.5}$) (see Table 1). Gomot and Pihan (1997) quantified variable affinities of snails for Cu by using bioaccumulation factors (BF) from ambient environment to organisms or contamination factors (CF) related to accumulation in reference animals (Table 2). These two factors are important, because the BF provides a measure of the degree to which land snails take up and retain Cu from all exposure routes, whereas CF reveals the level of contamination that results from Cu exposure.

2.1 Exposure via Air

Inhaling tiny subdivisions of solid matter that results from pollution and are suspended in a gas or liquid (particulate matter or PM) is known to produce health hazards (Davidson et al. 2005; Sacks et al. 2011). Among the PM that poses the highest threat are those particles referred to as PM_{10} (particulate matter with a diameter <10 μm) and $PM_{2.5}$ (particulate matter with a diameter <2.5 μm); these particle sizes are known to produce higher metabolic risk to animals (including terrestrial pulmonates) via respiration (Vinitketkumnuen et al. 2002). The toxic hazards imposed by atmospheric contamination of copper is enhanced relative to other paths because Cu particles may be easily resuspended in atmosphere as fine dust after they are removed by gravitational settling, dry disposition, rain, or snow (Barceloux 1999).

Nonferrous metal processing through pyrometallurgically or hydrometallurgically methods represents a major source of copper emissions. Thus, Jorquera (2009) reported higher PM concentrations than the standard ones (Table 1) in a copper mining area near Tocopile (Chile), wherein PM_{10} ranged between 79 and 117 μg/m³; in contrast, $PM_{2.5}$ reached a concentration of 20 μg/m³. These values decreased progressively as distance from the source increased, as shown by Sánchez-Rodas et al. (2007), Sánchez de la Campa et al. (2008), and Zhang et al. (2008). Other significant sources of copper aerial pollution were associated with urban areas and resulted from

Table 2 Selected laboratory data on the tolerance of land snails to copper, in the context of age, species, chemical form, and duration of exposure

Chemical form	Species/age	Source and time of exposure/snail origin	Contaminant dose	Biomarker of copper exposure	Parameter	Reference
$CuCl_2$	Helix aspersa (3-month-old)	Food/90 days (breeding)	19–243 ppm	Bioaccumulation factor (foot)	BF=1.60–5.70	Gomot and Pihan (1997)
				Bioaccumulation factor (viscera)	BF=1.80–3.30	
	Helix aspersa Maxima (3-month-old)			Bioaccumulation factor (foot)	BF=2.20–9.00	
				Contamination factor (foot)	CF=0.59–29.30	
				Bioaccumulation factor (foot)	BF=2.10–5.30	
				Contamination factor (foot)	CF=0.95–21.04	
	Helix aspersa (1 month, 1 g)	Food/28 days (indoor breeding)	100–4,000 ppm	Rate of food consumption	EC_{50} = 1,180 ppm NOEC = 500 ppm	Gomot de Vaufleury (2000)
$Cu(NO_3)_2$	Helix aspersa (8-month-old)	Food/90 days (indoor breeding)	16–1,789 ppm	Rate of food consumption	EC_{50} = 1,350 ppm EC_{20} = 248 ppm	Laskowski and Hopkin (1996a)
	Helix aspersa (adult)	Food/120 days (indoor breeding)			EC_{50} = 859 ppm EC_{20} = 275 ppm	
				Clutch size	EC_{20} = 533 ppm EC_{50} = 1,050 ppm	
				Copper accumulation (soft tissues)	Control: 101 ± 35 ppm Treatment: 191–906 ppm	Laskowski and Hopkin (1996b)
				Copper accumulation (shell)	Control: 0.07 ± 0.03 ppm Treatment: 0.09–4.80 ppm	

(continued)

Table 2 (continued)

Chemical form	Species/age	Source and time of exposure/snail origin	Contaminant dose	Biomarker of copper exposure	Parameter	Reference
Cu(CH3COO)$_2$	*Theba pisana* (weight: 1.24±0.012 g, shell diameter: 17.3±0.025 mm)	Food/5 weeks (nature)	50–15,000 ppm	Rate of food consumption	EC_{50} = 979 ppm (2 weeks) EC_{50} = 522 ppm (3 weeks) EC_{50} = 239 ppm (4 weeks) EC_{50} = 56 ppm (5 weeks)	El-Gendy et al. (2011)
				Increase in total fresh weight	EC_{50} = 8,418 ppm (2 weeks) EC_{50} = 6,946 ppm (3 weeks) EC_{50} = 6,462 ppm (4 weeks) EC_{50} = 4,961 ppm (5 weeks)	
		Topic contact/48 h (nature)		Mortality rate	LD_{50} = 37.88 ppm NOEC = 50 ppm LOEC = 10 ppm	Radwan et al. (2010)
CuSO$_4$	*Helix engadensis* (adult)	Food/28 days (nature)	4.00–2,500 ppm	Rate of food consumption	NOEC = 20 ppm LOEC = 100 ppm	Swaileh and Ezzughayyar (2000)
	Theba pisana (adult)	Food/48 h (nature)	10.61–21.22 ppm	Mortality rate	LD_{50} = 26.54 ppm	El-Gendy et al. (2009)
	Archachatina marginata (55–75 g)	Food/7 days (nature)	Food was sprinkled with Cu solution, 2.00–3.20 mM	Mortality rate	LC_{50} = 2.03–2.72 mM LC_{95} = 2.45–4.25 mM	Otitoloju et al. (2009)

Note: *BF* bioaccumulation factor, *CF* contamination factor, *EC* half maximal effective concentration, *NOEC* no observed effect concentration, EC^{20} effective concentration at 20% of the maximum response, *LOEC* lowest observed effect concentration, LC^{50} median lethal concentration, LC_{95} lethal concentration resulting in 95% mortality several combustion processes to wit: coal-fired power stations, vehicular traffic, and waste incinerators (Bertram et al. 2002; Yoo et al. 2002).

Since many land snails are sinantropic species (e.g., *Helix aspersa*) that breathe through their lungs as humans do, it is expected that they may efficiently mirror the air quality in large urban areas (Gomot de Vaufleury and Pihan 2000). For example, the Russian company Vodokanal uses six Giant African snails (*Achatina fulica*) to monitor pollutants exhausted from a sewage incinerator (Titova 2011). Regoli et al. (2006) reported mortality rates <10% for the gastropod *Helix aspersa* (4–6 g total weight) after 4 weeks of exposure to exhaust gases from vehicular traffic. These snails had been settled in plastic cages and did not have soil contact; they were fed on uncontaminated vegetables and were deployed within 1 m of the road margin in several Italian sites. The average concentration of trace copper in the midgut glands of these animals, as assessed by atomic absorption spectrophotometry with electrothermal and flame atomization, ranged from 8.65 ± 1.34 up to 80.80 ± 22.90 ppm dry wt. Hence, from these aforementioned examples it is obvious that there are objective scientific data that sustain the concept of using land snails as bioindicators of pollutant emissions.

2.2 Exposure via Soil

In soils, copper exists as both insoluble species and soluble ones bound to anions such as chloride (Cl^-) or sulfate (SO_4^{2-}) (Sauvé et al. 1997). Although soil accumulation of Cu varies by pH and organic matter content (Jeffery and Uren 1983; Landner and Lindestrom 1999), free amounts are insignificant compared to the system's total copper burden (Sauvé et al. 1998). Once deposited in soil, Cu remains relatively immobile over long periods of time (Nanda Kumar et al. 1995). Hence, soil-bound Cu poses a higher risk to the health of land snails as compared to uptake via inhaled air. The soil properties influence metal accumulation in land snails (Gimbert et al. 2006). Pauget et al. (2011) recently revealed, for the first time, the key parameters that affect metal transfer from soil to terrestrial mollusks (*Helix aspersa*). The results showed that soil alkalinization and level of organic matter content caused decreased metal (Cd, Pb) bioavailability to *Helix aspersa*, whereas Kaolin clay had no significant effect.

Harrison et al. (1999) reported that the maximal concentrations of copper in soil that are protective to soil-residing organisms should range between 40 and 100 ppm, the exact level depending on the soil type. Elevated soil copper levels are associated primarily with copper mining, smelting, and other copper-associated industries (Perez and Calvo de Anta 1992; Kabala and Singh 2000; Mighall et al. 2002); soil contamination from such activities was restricted primarily to soil surface horizons and was directly related to the distance from the mining and smelting areas (Kabala and Singh 2000). In addition, high soil concentrations of copper occur in areas where waste from sewage treatment plants are disposed of onto soil or are used as agricultural fertilizers (Singh et al. 2004). Abuse or over use of copper as an additive in animal nutrition also adds to the Cu soil levels from the use of manure as a natural fertilizer (Ogiyama et al. 2005). Furthermore, rather high concentrations of total copper (ranging from 38 up to 251 ppm) have been found in the upper soil horizon of vineyards (Brun et al. 2001; Pietrzak and McPhail 2004; Rusjan et al. 2007; Druart et al. 2011). Such findings are extremely important, because land snails come in contact with these surface horizons, especially through their motility, egg-laying, and feeding activities in soil (Gomot et al. 1989; Heller 2001) and as a consequence face potential threats to their health.

Terrestrial gastropods regularly eat soil (Gomot et al. 1989; De Grisse et al. 1996; Elmslie 1998; Coeurdassier et al. 2002), and therefore, they bioaccumulate copper not only via food from the food chain but also via contaminated soil. Most studies address the bioaccumulation of heavy metals along trophic chains via food uptake and do not address direct uptake from soil by cutaneous contact and/or by eating contaminated soil. As a result, little information exists concerning direct copper accumulation from soil. Coeurdassier et al. (2002) used an original approach to assess the contribution of digestive and epithelial transfer of Cd from soil to *Helix aspersa*. In this study, the snails were kept either in direct contact with the soil or separated from the substrate with a perforated plate, i.e., allowing the ingestion of soil, but avoiding the epithelial contact. Interestingly, the results showed that cutaneous and digestive routes of exposure are equally efficient in transferring Cd from soil to snail.

Because the foot of land snails comes into direct contact with contaminated soils, the epithelial transfer is expected to occur as an important additional path of exposure. Severe signs of toxicity have been reported to adult *Helix aspersa* exposed to solutions of 0.01–0.1 ppm copper sulfate via the epithelium; such toxicity was directly related to snail mortalities, the dose rate of the metal, and the amount of mucus exuded (Bonnelly de Calventi 1965). Ryder and Bowen (1977a) demonstrated cutaneous uptake of copper into the foot of slug *Agriolimax reticulatus*. The slugs crawled on filter paper saturated with 1,000 ppm copper sulfate solution for 30 min. It was found that endocytosis is the primary mechanism by which cutaneous transfer of molecules into the gastropod foot occurs (Ryder and Bowen 1977b).

Several studies have also shown that transporting proteins for different ions are expressed in snail epithelial cells (Robertson 1964; Ahearn et al. 1994; Pivarov and Drozdowa 2002). Among other functions, the epithelium of snail foot regulates the quantity of water in the mucus, excretion of metabolites, and transepithelial ion fluxes (Machin 1977; Simkiss 1988; Luchtel and Deyrup-Olsen 2001). Therefore, the epithelial transfer of Cu ions may represent an important penetration route for highly soluble copper salts, such as copper sulfate. In addition, evidence suggests that other possible sources of Cu buildup in living snails may exist than simple contact with soil residues or vegetation consumption. For example, it was found that the earthworm *Aporrectodea tuberculata* increased the soil bioavailability of Cu to *Helix aspersa* (Coeurdassier et al. 2007). As a result, future studies must be designed to elucidate the physiological mechanisms specific to this route of contamination and its potential importance in copper transfer along food chains.

2.3 Exposure via Food

Most researchers investigating Cu exposure via food have conducted controlled environment laboratory studies, in which test snails were fed fodder enriched with selected levels of copper. Exposure to such Cu-enriched foods inhibited land snail (*Theba pisana*) feeding and growth in a dose-dependent manner (El-Gendy et al. 2011). Similar effects were reported for other Cu-exposed terrestrial snails such as *Helix pomatia* (Moser and Wieser 1979), *Helix engaddensis* (Swaileh and Ezzughayyar 2000), *Helix aspersa* (Gomot de Vaufleury 2000), *Archachatina marginata* (Otitoloju et al. 2009), or *Limicolaria flammea* (Amusan et al. 2002). The degree of copper toxicity suffered by snails depends on various factors, such as snail species, chemical form of the Cu, diet composition (Gomot and Pihan 1997), time of exposure, age, size (Laskowski and Hopkin 1996a), or physiological status (Moser and Wieser 1979). Regardless of the foregoing factors, the most significant effects of Cu toxicity were observed to occur at the highest concentrations, and minimal or no effects were observed at low and intermediate doses (as shown in Table 2).

In a study with *Helix pomatia*, about 97% of the ingested Cu accumulated in the snail body, and the copper assimilation efficiency remained high as long as food ingestion did not exceed 3% of the fresh snail body wt per day (Moser and Wieser 1979). It was found that this species was able to efficiently absorb copper from their diet (Dallinger and Wieser 1984a) and was able to maintain a relatively stable Cu concentration in their tissues (Dallinger et al. 2000). Similarly, *Arianta arbustorum* fed on copper-enriched agar plates had an assimilation rate which always exceeded 95% (Berger and Dallinger 1989). In addition, the copper content of *Levantina hierosylima* was strongly positively correlated with body wt ($r^2 = 0.96$), showing that Cu levels were dependent on body wt (Swaileh et al. 2001). One plausible explanation for such high assimilation rates may be the fact that in natural environments Cu always occurs at concentrations near the minimum nutritional requirements of invertebrate (Hopkin 1993a, b).

Scheifler et al. (2006) investigated the transfer of Cu in a soil–plant (lettuce, *Lactuca sativa*)–invertebrate (snail, *Helix aspersa*) food chain under laboratory conditions. The snails were exposed for 8 weeks to Cu either via contaminated lettuce or via contaminated soil and lettuce. It was found that Cu concentrations did not increase during any of these experiments. Although incomplete, these data provides the basic argument that copper does not biomagnify along such food chains, but it is rather efficiently regulated in the land snail body.

2.4 Field Exposure

In natural environments, land snails (*Pulmonata*) are exposed simultaneously to heavy metals through multiple routes of exposure (Notten et al. 2005; Gomot de Vaufleury et al. 2006). Therefore, Cu uptake may be regarded as a cumulative process that occurs via mixed air, soil, and food exposures, i.e., field exposure. Such findings are extremely important, because terrestrial pulmonates spend their entire lives on or in the upper soil horizons, and in polluted areas they routinely consume contaminated soil, vegetation, and organic debris. Jones (1991) surveyed the area near the Avonmouth nonferrous smelting works (in the UK) and reported the absence of *Helix aspersa*, as a result of the extremely high levels of Cu pollution near the factory. Other investigations (Forray 2002) similarly showed the absence of land snail *Cepaea vindobonensis* near the Zlatna copper ore-processing and smelter plant (Alba County, Romania).

The native land snail, *Levantina hierosylima,* sampled from several locations in the West Bank—Palestine (Jerusalem, Abu-Dies, Qarawa, and Taibeh) revealed an average copper concentration of 126.90 ppm, with significant differences in concentration among the four sites (Swaileh et al. 2001). The concentration differences were attributed primarily to site variability in atmospheric fallout from traffic roads and industrial facilities.

Druart et al. (2011) investigated the reliability of *Helix aspersa* as a bioindicator of pesticide drift, deposit, transfer, and effects in a vineyard near Bergbieten (Bas Rhin, France). The juvenile snails (4–7 g total wt) were caged for 10 days and exposed to Bordeaux mixture ($CuSO_4$ 7.6 g/L) via contaminated soil and wild vegetation. The results revealed an increase of copper concentration along the food chain from 41.1 ppm Cu in the soil up to 140 ppm Cu in the snail body (foot and viscera). The copper content in the ingested vegetation, foot, and viscera of *Helix pomatia* (reared in open air farms) respectively were 1.79, 11.81, and 18.01 ppm (Jokanović et al. 2006). Additional data concerning Cu accumulation in land snails under field conditions are shown in Table 3.

Cu concentrations naturally occurring in plants generally do not exceed 10 ppm, and the typical ratio between Cu concentrations in plants vs. soil has been estimated to be 0.25 (Payne et al. 1988). To be nontoxic to plants, the maximum normal Cu concentration must not exceed 50 ppm (Davis and Beckett 1978; Gogoasa et al. 2011). Phytoremediation studies in polluted areas have revealed that the copper content of vegetation varies widely among different species, with dry wt vegetation levels ranging from 0.6 ppm in *Agropyron* sp. to 80.8 ppm in *Artemisia* sp. (Porębska and Ostrowska 1999). The Cu levels in gastropod food are directly related to which plant types are preferred and the degree to which Cu is accumulated in these preferred plants. Studies conducted by Wolda et al. (1971), Carter et al. (1979), Lazaridou-Dimitriadou and Kattoulas (1991), Desbuquois and Daguzan (1995), Iglesias and Castillejo (1999), and Chevalier et al. (2001) revealed that snails preferred such plants as nettle (*Urtica dioica*), lettuce (*Lactuca Sativa*), dandelion (*Taraxacum officinale*), or members of the *Poaceae* family. Moreover, nettle (*Urtica dioica*) and dandelion (*Taraxacum officinale*) have proved to be reliable plant indicators of copper contamination (Królak 2003; Yildiz et al. 2010).

Because land snails are major herbivores in natural terrestrial ecosystems (Beeby 1985), Cu exposure via food uptake is expected to serve as the main path of their contamination. The best way to assess land snail validity, reliability, and viability as bioindicator of Cu pollution under field conditions is to relate copper concentrations

Table 3 Selected data on the accumulation of copper in land snails under field conditions, in the context of species, origin, age, body weight, and analyzed organ

Species	Origin	Study purpose	Details on snails and/or sampling area	Analyzed organ	Copper level (ppm)	Reference
Helix aspersa aspersa	Indoor breeding		Mean weight of snails, 2.5 or 3.5 g	Body without shell	85–162	Clayes and Demeyer (1986)
			Estimated age, 3 months	Foot	179–619	Gomot and Pihan (1997)
				Viscera	157–422	
	Nature	Passive biomonitoring	Snail of all sizes, uncontaminated and contaminated sites	Body without shell	57–534	Beeby and Richmond (2003)
				Hepatopancreas	27–568.3	
				Soft tissues without hepatopancreas	61.9–474.2	
	Indoor breeding	Active biomonitoring	Mean weight of snails, 4–6 g	Hepatopancreas	7.31–103.7	Regoli et al. (2006)
			Mean weight of snails, 4–7 g	Foot, hepatopancreas	Max. value ≈145	Druart et al. (2011)
Helix aspersa maxima	Indoor breeding		Estimated age, 3 months	Foot	147 and 193	Gomot and Pihan (1997)
				Viscera	101 and 172	
			Adult snails	Foot	50.9	Łysak et al. (2000)
Helix pomatia	Nature	Passive biomonitoring	Uncontaminated and contaminated sites	Hepatopancreas	143 and 145	Pihan et al. (1994)
				Foot	172 and 135	
	Nature	Passive biomonitoring	Uncontaminated and contaminated sites	Hepatopancreas	139.4–802	Gheoca and Gheoca (2005)
				Foot	38.7–67.87	
				Mantle	49.8–86	
				Shell	4.95–8.74	
				Viscera	74.3–125.5	
Cepaea nemoralis	Nature	Passive biomonitoring	Contaminated sites	Foot	81	Teofilova et al. (2011)
			Different distances from a smelting site	Body without shell	30–86	Martin and Coughtrey (1982)
			Uncontaminated and contaminated sites	Body without shell	Max. value 200	Notten et al. (2005)
Arianta arbustorum	Nature	Passive biomonitoring	Uncontaminated and contaminated sites	Body without shell	100–400	Berger and Dallinger (1993)
Arianta arbustorum Bradybaena fructicum Aegonis verticillum	Nature	Passive biomonitoring	Uncontaminated and contaminated sites	Foot, hepatopancreas, kidney	Max. value 600 in kidney for *A. verticillum*	Rabitsch (1996)

in the gastropod body with concentrations in plants on which the snails feed. Several study authors have revealed the selectivity in food choice exhibited by many species of land snails (e.g., Baur et al. 1994; Hanley et al. 1995; Linhart and Thompson 1995). Therefore, to ensure accomplishment of the aforementioned task the food items that snails had consumed before being collected must be precisely identified (Dallinger et al. 2001a).

Several authors have studied the relationships between copper accumulation in soil, plant leaves, and snails. Notten et al. (2005) investigated Cu transfer in a soil–plant–snail food chain in the Biesboch area (The Netherlands). Cu accumulation was measured in soil, *Urtica dioica* leaves, and *Cepaea nemoralis* soft tissues (i.e., whole body without shell). Copper in the snail body was independent of the concentration in soil ($r^2 = 0.02$), but correlated moderately with Cu level in nettle leaves ($r^2 = 0.39$). By contrast, Nica et al. (2012) reported a strong correlation between Cu content in *Helix pomatia* foot and soil ($r = 0.90$). However, the amount of Cu present in the snail body (i.e., foot, hepatopancreas) was generally independent of concentrations in soil and nettle leaves ($r = 0.20$–0.62, $p > 0.05$). In addition, Beeby and Richmond (2003) found no relationship between copper levels in *Helix aspersa* soft tissues (i.e., whole body, whole body without hepatopancreas, hepatopancreas) and concentrations in both vegetation (dandelion—*Taraxacum officinale*) and soil ($r = 0.02$–0.08). In summary, these data reveal that land snails do not serve as a "sink" for copper accumulation along terrestrial food chains, and moreover, they suggest that these mollusks are metabolically able to tolerate and regulate external copper influx.

3   Metabolic Regulation of Copper in Land Snails (*Pulmonata*)

Unlike the hemoglobin in red blood cells found in vertebrates, hemocyanins are respiratory metalloproteins, which are suspended directly in the gastropods' hemolymph (Mikkelsen and Weber 1992) and contain two copper atoms that reversibly bind a single oxygen molecule (Velkova et al. 2010). As a result, all pulmonate species, including the terrestrial ones, have blue rather than red blood (when oxygenated), excepting a single gastropod family, the *Planorbidae* (Markl 1986). Although these proteins are, at most, one-fourth as efficient as hemoglobin at transporting oxygen, their free-floating nature allows them to carry higher amounts of oxygen at higher densities (Karlin et al. 1987). In addition, copper peptides and superoxide dismutase (SOD) are naturally found inside the complex glycoconjugates secreted by land snails when under stress (Siddiqui et al. 2009).

Studies conducted on *Helix pomatia* by Moser and Wieser (1979) revealed basic information about how copper is metabolized, stored, or transmitted during digestion. Ingestion of copper-enriched foods initially induced a rapid increase of the Cu concentration in the anterior alimentary tract. This was followed 3 days later by a massive Cu level increase in the midgut glands, and finally, after 5 days, by a further accumulation in the posterior part of the alimentary tract, especially in the intestine.

A major part (97%) of copper ingested via contaminated lettuce ($1,390 \pm 420$ ppm) remained in the snail body, with the midgut and the albumen glands being the main storage sites for bound copper. In addition, the amounts of copper appearing in feces was independent of the food type ingested, and Cu levels found were not so different from those found in feces of organisms ingesting untreated lettuce.

Swaileh and Ezzughayyar (2000) showed to *Helix engaddensis* that the inhibitory effect of consuming a Cu-contaminated diet is reversible after the end of an exposure event. Hence, a few days after cessation of copper administration it was quickly eliminated from the body, and intratissular levels decreased to normal values as observed on other snail species by Dallinger (1996). Sulfur-reducing bacteria that exist in the gastropod esophageal crop can facilitate Cu absorption in the snail body (Simkiss 1985). Taken together existing data on how land snails accumulate, excrete, and metabolize Cu, show that these organisms are a class of invertebrates that are able to efficiently regulate copper influxes. How snails metabolically respond to Cu uptake also reinforces the thought encompassed by the last sentence.

Achatina fulica exposed to sublethal doses of copper displayed differential responses of tissue carbohydrate level and phosphatase activity; these results pointed to the lactate/pyruvate ratio and intratissular calcium content as direct evidence for the snail adaptation to Cu-induced toxic stress (Ramalingam and Indra 2002). After 10–12 weeks of exposure, fecal copper levels reached their peak, whereas the total excreted copper progressively increased (Ireland and Marigomez 1992). Feces as a main excretory path for Cu was affirmed in other studies as well (Moser and Wieser 1979; Laskowski and Hopkin 1996a; Zödl and Wittmann 2003). In contrast, there is no evidence to support excretion of copper via the urine, although mucus is thought to be a potential route of heavy metal excretion (Menta and Parisi 2001).

Feeding experiments have revealed that most ingested copper remains in the snail body, and little is excreted via feces, urine, or mucus vs. the total ingested amount. Therefore, one can conclude that copper is metabolically regulated through other physiological mechanisms. Several researchers found that the accumulation and regulation of Cu is associated particularly with tissues of the mantle and hepatopancreas (e.g., Berger and Dallinger 1989; Laskowski and Hopkin 1996b; Snyman et al. 2000). The latter organ was regarded as central to the metabolism of land gastropods, and among other functions acts as the main site of metal storage and detoxification (Janssen 1985). The metabolic pathway that trace elements follow inside mollusks' bodies is dependent on the physicochemical properties of the element involved. Copper is regarded to be among the so-called border-line metals (class C) that include trace elements that can bind both oxygen- and sulfur-bearing ligands (Nieboer and Richardson 1980). Once absorbed, Cu may follow the same pathway by which calcium is metabolized, and therefore, it may finally precipitate within insoluble calcium granules from hepatopancreatic basophilic cells (Simkiss 1981; Simkiss et al. 1982; Almendros and Porcel 1992). Several study authors reported that copper precipitation occurred in this manner to *Arion ater* (Ireland 1979; Marigòmez et al. 1986). In some gastropod species, these granules are released into the hepatopancreatic tubular lumen, and as a consequence, may provide an additional route of trace element excretion (Simkiss and Mason 1983).

Several authors have revealed that, in addition to cellular compartmentalization, heavy metals (e.g., Cu, Zn, Cd, Hg, and Ag) can be sequestered in the snail tissues by complexation to specific metallothioneins (MTs). These proteins, which are located on Golgi apparatus membrane, lower heavy metal availability within cells, and hence, reduce their cytotoxic potential (Kägi and Schäffer 1988; Dallinger 1996; Sigel et al. 2009). In addition, MTs may also be involved in regulating Zn and Cu homeostasis (Cherian and Goyer 1978).

Berger et al. (1997) was the first to characterize a specific Cu metallothionein (HpCuMT), which in their case, was from the mantle tissue of *Helix pomatia*. This protein, coded in Protein Knowledgebase as P55947 (MTCU_HELPO), consists of 64 amino acids, of which 14 cysteine residues are arranged in C-X-C groups (UniProtKB 2011). Although the HpCuMt isoform presented the same arrangement of cysteine like the specific cadmium metallothionein isoform (HpCdMT) obtained from *H. pomatia* midgut glands, these two MTs were different in total length and at 26 positions along their peptide chains (Dallinger et al. 1997, 2004b). Because the thiol group of this protein's cysteine residues allows it to carry copper (I) ions, the HpCuMT isoform was associated with regulating Cu in the context of the synthesis and metabolism of hemocyanin (Dallinger et al. 2001b; Chabicovsky et al. 2003). Similarly, it was demonstrated that the CaCuMT isoform, isolated from *H. aspersa*, is deeply involved in Cu homeostasis, sharing about 30–50% of the task of maintaining Cu balance (Höckner et al. 2011).

Recently, the aforementioned role of HpCuMT isoform in hemocyanin synthesis and metabolism has been confirmed. Thus, it was demonstrated that the copper in the Roman snail (*Helix pomatia*) is regulated only by the rhogocytes (Dallinger et al. 2005), which are specialized cells that occur only in mollusks (Albrecht et al. 2001; Sturm et al. 2006). Rhogocytes appear as large cells (up to 30 µm long) which are either scattered singly throughout the connective tissue of body parts (particularly in mantle, foot, gut, and digestive glands) or aggregated in small cell accumulations placed around blood the sinus and/or blood vessels (Dallinger et al. 2005).

Notwithstanding the foregoing, the precise mechanisms involved in the metabolic regulation of copper in land snails (*Pulmonata*) are still unknown. It was found that both whole rhogocytes and their storage products can be eliminated from the bodies of land snails by diapedesis (Brown 1967; Marigomez et al. 2002). Nevertheless, rhogocytes seem to possess endocytotic and phagocytotic abilities (Haszprunar 1996), and their potential role in metal uptake, intracellular trafficking, and extrusion out of land snails cells cannot be excluded. In addition, Dallinger et al. (2005) suggested that specific carrier proteins may be involved in the efficient regulation of copper in terrestrial pulmonates. In fact, similar mechanisms were found to exist in yeasts (Riggio et al. 2002) and in mammals (Pase et al. 2004). Moreover, Fisker et al. (2011) revealed the possible existence of a genetic component in the inheritance of copper tolerance to other soil invertebrates (i.e., earthworm *Dendrobaena octaedra*). Similar mechanisms are likely to exist in terrestrial gastropods (*Pulmonata*), and therefore, in the future, researchers must further investigate and clarify the physiogenetic background of the land snail's ability to adapt and live in Cu-polluted areas.

4 Bioaccumulation of Cu by Land Snails

Because many species of land snails (*Pulmonata*) are able to concentrate trace metals in their tissues, these mollusks can be included among the so-called "bioindicators of accumulation" (Dallinger et al. 2001a). Dallinger (1994) defines the biological concentration factor as the slope (a) of linear relationship existing between metal concentration in the environment (x) and metal accumulation in the gastropod's body (y). Many land snail species (e.g., *Helix pomatia*, *Helix aspersa*, *Cepaea nemoralis*) accumulate Cu far above environmental concentrations ($a > 2$), and therefore, they function as "macroconcentrator" species for Cu (Dallinger and Rainbow 1993; Laskowski and Hopkin 1996a).

Although copper is widespread throughout land snails tissues, it does not accumulate at specific locations like lead or cadmium do (Coughtrey and Martin 1976). Most authors reported that Cu distribution among body organs is particularly associated with the snail foot, mantle, and midgut glands/hepatopancreas. This trend (see Tables 2 and 3) is valid for most species of terrestrial gastropods such as *Helix pomatia* (Moser and Wieser 1979; Dallinger and Wieser 1984b; Gheoca and Gheoca 2005), *Helix aspersa aspersa* (Laskowski and Hopkin 1996a; Gomot and Pihan 1997; Druart et al. 2010), *Helix aspersa maxima* (Gomot and Pihan 1997), *Arianta arbustorum* (Berger and Dallinger 1989; Rabitsch 1996), *Achatina fulica* (Ireland and Marigomez 1992), *Aegonis verticillus*, *Bradybaena fructicum* (Rabitsch 1996), and *Archachatina marginata* (Otitoloju et al. 2009).

Adeyeye (1996) and Özogul et al. (2005) reported copper accumulation in the meat of edible snails, but did not precisely identify the source of the Cu, which may have been from natural occurrence or from contamination through ingestion of polluted soil, food, or organic debris. Interestingly, quite different Cu concentrations were reported to exist in the foot of contaminated snails. Moser and Wieser (1979) found Cu concentrations from 133 ± 36 up to 817 ± 55 ppm dry wt in the foot of Cu-exposed *Helix pomatia*, whereas Gomot and Pihan (1997) determined higher values that ranged from 1,261.90 to 2,113.00 ppm dry wt in the foot of Cu-contaminated *Helix aspersa*. Both species were reared under laboratory conditions and were fed Cu-enriched diets. In addition, several authors have revealed the levels to which copper bioaccumulates in different land snail species (Berger and Dallinger 1993; Rabitsch 1996), and therefore, one can conclude that bioaccumulation levels are related to their different ecophysiological particularities. Because even within closely related species of land snails there are significant differences in copper uptake, accumulation, and detoxification (Gomot and Pihan 1997), the comparisons of concentrations between different species must be treated with caution, or, preferably, avoided.

Since land snail's nutrition may be compromised by heavy metal pollution, it is expected that the copper content of land snails will range widely, even within the same species. Babafola and Akinsoyinu (2009) reported that *Limicolaria* sp. and *Achatina* sp., sampled from Ibadan (Ido State, Nigeria), contained copper concentrations in the foot that were between 0.29 ± 0.07 and 1.03 ± 0.07 mg/100 g of fresh meat, whereas Fagbuaro et al. (2006) detected no trace of copper in snails purchased from Ado Ekiti (Ekiti State, Nigeria). Gomot and Pihan (1997) showed that the bioaccumulation capacity is higher for the smaller subspecies *Helix aspersa minima* (syn. *Helix aspersa*) vs. *Helix aspersa maxima* (Table 2). This seems paradoxical, but it might be associated with different digestive physiology between the species, as proposed by the latter authors, rather than from differences in food composition, as suggested by Greville and Morgan (1990), or from chemical speciation of Cu in the food of snails, as inferred by Berger and Dallinger (1989). In addition, the amount of copper per unit weight of tissue increased with size and age within the same species (Marks 1938). Gomot and Pihan (1997) thought this was from the more intense metabolism that occurred in growing animals. Moreover, Coughtrey and Martin (1977) found that the copper level increased in a linear fashion with *H. aspersa* weight. Because winter dormancy is associated with the cessation of hepatopancreatic detoxifying activity, higher rates of copper bioaccumulation are expected to occur in hibernating snails than in the active ones. For example, Moser and Wieser (1979) revealed that adult *H. pomatia* fed for 3 weeks with copper-enriched lettuce retained Cu at values up to eight times higher during the winter than during the summer. Moreover, *H. aspersa* exhibited better uptake efficiency of copper (II) sulfate than copper (II) chloride (Gomot and Pihan 1997). Therefore, one can conclude that copper accumulation in the snail body may be affected by both genetic and physiological factors.

Most studies on accumulation of Cu in land snails (*Pulmonata*) have shown that the animals usually tend to regulate this metal content in their soft tissues, rather than concentrating it as a function of environmental Cu concentrations in the soil (see Sect. 2.4). Hence, any increase of Cu concentrations in the land snail organs, in response to external Cu uptake, have generally proved to be transient (e.g., Moser and Wieser 1979; Swaileh and Ezzughayyar 2000), whereas an important fraction of Cu in snail soft tissues was always bound to a Cu-specific MT isoform, irrespective of environmental exposure (Berger et al. 1997). Berger and Dallinger (1993), Gomot and Pihan (1997), and Dallinger et al. (2004a) showed that copper bioaccumulated ten times higher in the tissues of snails living in polluted soil habitats (10^{-3} to 10^{-2} mol/kg dry mass) than it did in snails inhabiting unpolluted areas (10^{-4} to 10^{-3} mol/kg dry mass). Dallinger et al. (2005) reported that Cu-exposed *Helix pomatia* concentrated only two to three times higher amounts of copper in their soft tissues than did the control animals. The Cu-exposed snails were fed with Cu-enriched lettuce leaves (7.22 ± 1.38 μM Cu/g dry wt) for 15 days, whereas the control animals were kept under the same conditions but were fed with uncontaminated food (0.35 ± 0.13 μM Cu/g dry wt). This moderate pattern of copper bioaccumulation in the snail soft tissues is in accordance with land snail ability "to maintain Cu concentrations in their organs within a moderate range, even under conditions of increased Cu availability in the environment" (Dallinger et al. 2000). These results are in contrast to the long-term accumulation of nonessential metals (e.g., Cd) by these animals and do favor the use of terrestrial snails (*Pulmonata*) as a long-term Cu bioindicator species.

Although, other divalent metals such as cadmium (Beeby and Richmond 1987) or lead (Beeby and Richmond 2002) do accumulate in the shells of terrestrial pulmonates, little information exists regarding copper accumulation in this organ.

However, there are premises to assume that copper may accumulate in snail shells for a longer time than in their soft tissues. Traces of copper were found in the shells of *Cepaea vindobonensis* that were sampled both from polluted and unpolluted areas of the southwestern Apuseni Mountains (Forray 2002). Because the gastropod blood contains hemocyanin, it is expected that shells will contain at least traces of Cu. Copper accumulations of 0.10 up to 2.30 ppm dry mass were reported in the shells of *Archachatina marginata* and *Achatina fulica* that were sampled from different sites in Nigeria (Oyedepo et al. 2007; Aboho et al. 2009). Both *Archatina sp.* and *Limicolaria sp.* inhabiting the same area, however, exhibited higher copper concentrations in the shell (viz., 4.31 and 16.98 mg/L) (Jatto et al. 2010).

Laskowski and Hopkin (1996b) reported that the copper content in *Helix aspersa* shells increased with Cu exposure concentrations, but its accumulation (<3.70 ppm) was insignificant as compared to those in soft tissues (ranging from 66 up to 906 ppm). In contrast, Gbaruko and Friday (2007) found that *Archachatina marginata* had almost threefold higher concentrations of copper in the shell (2.10 ± 0.03 ppm) than in the body (0.89 ± 0.04 ppm dry mass). Because the duration of most studies was only a few months, snails were exposed to copper for only short time periods, which probably restricted significant transport into the shells (Laskowski and Hopkin 1996b). Although little information exists concerning the influence of Cu on shell traits, such information does exist for aquatic species. The mud snail *Cipangopaludina chinensis malleata* presented an inverse correlation between shell width and concentrations of copper in the snail body (Kurihara et al. 1987), whereas *Lymnaea stagnalis* deposited copper mostly in the shell (Pyatt et al. 1997). Contextually, copper may accumulate in the shell of animals exposed for all or the majority of their lives. As indirect proof of this, Eeva et al. (2010) found spatial variations in the shell mass of terrestrial snails that were indirectly associated with Cu levels near a copper smelter; shell size, and mass were highest in moderately polluted areas and smallest in the most polluted ones. Further studies are required to expand on these findings and to elucidate the physiological significance of these results.

5   Biomarkers of Copper Pollution in Land Snails

Biomarkers are defined as chemicals, metabolites, susceptibility characteristics, or changes in the body that relate to the exposure of an organism to a chemical or a group of chemical agents (Walker et al. 1996). Unlike bioindicators, biomarkers are not species but subindividual parameters (Dallinger et al. 2001). The presence of a biomarker does not necessarily reveal the source or route of exposure (Paustenbach and Galbraith 2006), but it can be used to determine if an exposure has occurred and the resulting effects of the exposure (Gil and Pla 2001). Although the cytochrome P450 monoxygenase system is a commonly used biomarker in mollusks as prognostic tools for increased levels of pollution (e.g., Peters et al. 1999; Livingstone et al. 2000; Michel et al. 2001; Zanette et al. 2010), little information exists concerning its engagement in pollutant metabolism by land snails (Ismert et al. 2002). Instead, several measures of exposure and effect can be used in these mollusks as biomarkers of copper exposure: (a) cellular biomarkers (ultrastructural and hispathological alterations); (b) lysosomal biomarkers (i.e., lysosomal stability, lysosomal membrane integrity); (c) induction of metallothioneins (MTs); (d) biomarkers of oxidative stress (i.e., balance between pro- and antioxidant factors, capacity of cellular antioxidants to neutralize reactive oxygen species—ROS, loss of DNA integrity). In Table 4 we provide data on the values of selected biomarkers in *Helix aspersa* that depend on copper dose and exposure time. This species was chosen not only because its physiology is well known (Gomot de Vaufleury 2000) but also because it is one of the most employed invertebrates for monitoring both accumulation and toxicologic effects induced by metal pollution (e.g., Laskowski and Hopkin 1996a, b; Gomot 1997, 1998; Snyman et al. 2000, 2009; Regoli et al. 2006).

5.1 Ultrastructural and Histopathological Alterations

Microscopy of accumulated Cu inside the gastropod body has indicated that Cu is deposited mainly in the foot, mantle, and midgut glands/hepatopancreas (Moser and Wieser 1979; Snyman 2001; Otitoloju et al. 2009). Smaller amounts are also present in the albumen glands, ovotestis, in connective tissue, and the kidney (Moser and Wieser 1979; Rabitsch 1996; Amaral et al. 2004). Copper exposure may be quantified by using atomic absorption spectrophotometry (AAS) methods. However, such analysis only provides quantitative information about accumulation (Lemos et al. 2009; Yetimoglu et al. 2010). In contrast, by assessing the ultrastructural and histopathological alterations that are induced at the cellular and tissular level (by using electron, fluorescence, or light microscopy; Jong-Brink et al. 1976) qualitative data is revealed concerning how land snails are physiologically able to regulate Cu in their bodies.

The digestive glands are the most important gastropod organs involved in pollutant detoxification (Klobu ar et al. 1997; Ismert et al. 2002). Gastropod epithelium consists of three cell types: excretory, calcium, and digestive cells, the latter of which is the most abundant cell type (Dimitriadis 2001). The digestive glands also possess connective tissue, which is composed of pigment cells, rhogocytes (pore cells), and fibroblasts among other tissue types (Amaral et al. 2004). Since the digestive glands are involved in sequestering and releasing metals (Ireland and Marigomez 1992; Soto et al. 1998), the resulting histological and histochemical changes are expected to be useful biomarkers of copper exposure (Snyman et al. 2009). For example, a 100-μM solution of copper (II) ions proved to be acutely toxic to the hepatopancreatic cells of *H. pomatia*. As a result, the viability of these cells was decreased (i.e., from 93% in controls to 87% in exposed groups), whereas bleb formation occurred at the cell membrane level (Manzl et al. 2004). *Arachachatina marginata* exposed to concentrations of up to 0.235 mM copper (II) sulfate revealed the prevalence of hepatocellular foci and hepatocytes that displayed peripheral thickening (Otitoloju et al. 2009). Adult *H. aspersa* fed a copper-contaminated diet exhibited significant effects of copper dose (i.e., exposure to 0, 80, and 240 ppm copper

Table 4 Selected values of cellular biomarkers in land snails exposed to copper under field conditions that are dependent on exposure route and duration, dose, and chemical form of the copper compound involved

Copper form/source of exposure	Contaminant dose (ppm)	Duration of exposure	Assay used for biomarker detection	Biomarker of copper exposure	Biomarker values	Reference
Copper oxychloride/food	Control (0)	6 weeks	Hematoxylin–eosin staining	Mean spermatozoan area (%)	65.23 ± 14.44	Snyman et al. (2009)
	80				71.79 ± 9.42	
	240	Control			69.65 ± 12.63	
	240	1 week			65.69 ± 8.26	
		2 months			64.39 ± 10.76	
					59.86 ± 21.96	
	Control (0)	6 weeks	Hematoxylin–eosin staining	Mean number of oocytes (µm²)	5.09 ± 3.44	
	80				3.52 ± 2.48	
	240	Control			2.33 ± 1.64	
	240	1 week			5.33 ± 3.58	
		2 months			5.02 ± 3.30	
					3.87 ± 2.80	
	Control (0)	Control	NRR assay	NRR time (min)	30.67 ± 6.33	
		3 weeks			23.11 ± 5.68	
		6 weeks			24.44 ± 8.55	
	80	Control			28.22 ± 5.33	
		3 weeks			16.00 ± 5.29	
		6 weeks			10.22 ± 3.53	
	240	Control			26.22 ± 6.04	
		3 weeks			12.89 ± 2.67	
		6 weeks			2.67 ± 2.83	

(continued)

Table 4 (continued)

Copper form/source of exposure	Contaminant dose (ppm)	Duration of exposure	Assay used for biomarker detection	Biomarker of copper exposure	Biomarker values	Reference
Airborne copper/air	8.65 ± 1.34	4 weeks	Spectrophotometry (240 nm)	CAT (μmol/min/mg protein)	321 ± 48.7	Regoli et al. (2006)
	21.2 ± 1.91				700 ± 144	
	80.8 ± 20.9				545 ± 51.3	
	8.65 ± 1.34		NADPH oxidation	GR (nmol/min/mg protein)	16.1 ± 3.54	
	21.2 ± 1.91				22.04 ± 5.08	
	80.8 ± 20.9				29 ± 7.77	
	8.65 ± 1.34		TOSC assay	TOSC toward ROO (U/mg protein)	981 ± 9.86	
	21.2 ± 1.91				1,588 ± 61.7	
	80.8 ± 20.9				1,330 ± 147	
	8.65 ± 1.34		TOSC assay	TOSC toward OH (U/mg protein)	1,071 ± 132	
	21.2 ± 1.91				1,544 ± 189	
	80.8 ± 20.9				1,206 ± 132	

Note: NRR assay neutral red retention assay, *NRR time* the retention time of the neutral red dye in the lysosomes, *CAT* catalase, *NADPH oxidation* oxidation of nicotinamide adenine dinucleotide phosphate-oxidase, *GR* glutathione reductase, *TOSC assay* total oxyradical scavenging capacity assay, *TOSC toward ROO* total oxyradical scavenging capacity toward peroxyl radicals (ROO), *TOSC toward OH* total oxyradical scavenging capacity toward hydroxyl radicals (OH) oxychloride for 6 weeks) and exposure time (i.e., 1 week and 2 months after exposure to 240 ppm copper oxychloride vs. control groups) on the height of hepatopancreatic epithelium cells and digestive glandular epithelium area, but not on the mean area of digestive gland tubules (Snyman et al. 2009). Use of autometallography disclosed that intracellular copper accumulations as visualized by light microscopy were black silver deposits and were associated with the relative volumetric density of connective tissue cells, but not with the relative volumetric density of calcium cells of the digestive gland epithelium (Amaral et al. 2004). In addition, under electron microscopy, rhogocytes and nucleus of snails that were not exposed to copper were entirely crowded toward the cell edge by a large central vacuole exhibiting a granular inner texture. By contrast, in the exposed snails, the large central vacuole was condensed and split into several smaller vacuoles that revealed an increased granule frequency as copper exposure increased (Dallinger et al. 2005).

It has been suggested in several studies that small amounts of copper may also be accumulated in snail cells other than those mentioned above (Triebskorn and Köhler 1996; Marigomez et al. 2002). One example of other tissues where such accumulations may be found is the reproductive organs (e.g., ovotestis, vesicula seminalis, and albumen glands). Otitoloju et al. (2009) reported that basophilic adenoma occurs within the *A. marginata* ovotestis when these organisms are exposed to a copper-enriched diet. Snyman et al. (2009) revealed significant time- and dose-dependent reductions in gamete densities in the ovotestis (i.e., spermatozoan area and oocytes number per 1 mm^2 ovotestis) (Table 4).

When exposed to concentrations of up to 0.1 ppm copper (II) sulfate via epithelial transfer, *Helix pomatia* revealed no alterations of mantle and foot mucopolysaccharides (Bonnelly de Calventi 1965). In contrast, *Achatina fulica*, fed for 4 weeks with copper-contaminated foods (i.e., 0, 180, 240, 320, 420, and 560 ppm Cu as copper sulfate), exhibited changes in total polysaccharide content of the albumen glands (Kalyani 1990). These polysaccharides are important constituents of perivitelline fluid, which provides nourishment to the embryo during development (Egonmwan 2008); such changes may therefore be associated with alterations of snail reproductive fitness. Thus, high Cu concentrations not only caused a sharp drop in snail fecundity (Snyman et al. 2009) but also induced the absence of oviposition (Laskowski and Hopkin 1996a).

Because assessing Cu-induced ultrastructural and histopathological alterations at the cellular level is expensive, complex, and time consuming, few such studies exist to complement biomonitoring results (Oehlmann and Schulte-Oehlmann 2002).

5.2   Lysosomal Stability and Membrane Integrity

Lysosomes are multifunctional cellular organelles that serve to digest different types of food molecules and foreign particles and to restore plasma membranes after cell injury (Blott and Griffiths 2002; Luzio et al. 2007; Tam et al. 2010). Molluskan haemocytes are able to sequester metals, including zinc, copper, and cadmium (George et al. 1978; McIntosh and Robinson 1999). Pollutant-induced stress causes the efflux of lysosomal hydrolases into cytosol (Regoli 1998). As a result of being based on these two features, the neutral red lysosomal retention assay (NRR) of haemocytes reflects physiological processes that occur after membrane damage and allow the assessment of haemocyte capacity in the context of adapting to stress conditions (Lowe and Pipe 1994). The digestion of neutral red cells in unbroken lysosomes produces an intense red coloring of these organelles, whereas any membrane damage interferes with the lysosomal ability to accumulate and retain the stain (Zhang et al. 1990; Regoli 2000). Because this assay has proved to be a rapid and sensitive method for evaluating lysosomal stability (Svendsen and Weeks 1995, 1997a, b; Svendsen et al. 2004), it has been extensively employed as a biomarker of mollusk exposure to environmental pollutants (e.g., Cajaraville et al. 1995; Donval and Plana 1997; Giamberini and Pihan 1997).

Snyman et al. (2000) was the first to use the NRR times of land snail haemocytes for assessing the influence of Cu exposure on lysosomal stability. Adult *H. aspersa*, fed a copper-contaminated diet, revealed that increasing Cu concentrations and times of exposure are associated with progressively shorter NRR times (Table 4). Similarly, for the same species, Regoli et al. (2006) reported an inverse relationship between internal Cu concentrations and NRR times. Itziou and Dimitriadis (2011) also found that Cu uptake via food has significantly decreased NRR times to *Eobania vermiculata*, both under field and laboratory conditions. The NRR assay can therefore serve as an useful cellular biomarker of land snail exposure to Cu compounds and particularly as an early warning system that can detect even low levels of metal contamination (Svendsen et al. 2004).

5.3 Induction of Specific Metallothioneins (MTs)

Because land snail Metallothioneins (MTs) are involved in heavy metal sequestration, these proteins are expected to function as biomarkers for risk assessment in terrestrial environments. The biochemical and functional features of MTs in land gastropods (i.e., strong inductibility, metal-binding specificity) may allow environmental scientists to assess the effect of metal pollution in time and space. As a measure of MT response to environmental pollution, researchers have relied on estimating either MT physicochemical properties (i.e., measurement of sulfyhydryl groups, immunochemical affinities, spectrophotometric absorption) or the metal content bound to MTs (i.e., direct quantification techniques, competitive metal displacement) (UNEP/RAMOGE 1999).

Berger et al. (1997) was the first to use a cadmium saturation method (viz., the cadmium-Chelex assay) to measure metallothionein concentrations in land snails. This method has allowed assessing not only protein saturation with cadmium but also the quantification of MT induction in digestive glands of Cd-exposed *H. pomatia*. Dallinger et al. (2004b) showed that for land snails (*Cepaea hortensis*) the cadmium binding of metallothionein (Cd-MT) is quickly responsive to cadmium exposures in a dose-dependent manner, and the inductive signal persists over long periods of time.

The Cd-Chelex assay has been modified by Dallinger et al. (2000) to assess MT concentrations (i.e., Cu-MT, total MT) in the mantle and the digestive glands of several helicid species. It was found that the Cu-MT level significantly increased in the digestive glands of Cu-exposed *H. pomatia*. Similar responses for total MT concentrations were reported for other Cu-exposed land snails, such as *Helix aspersa* (Regoli et al. 2006) or *Eobania vermiculata* (Itziou and Dimitriadis 2011).

Dallinger et al. (2005) revealed that, at the cellular level, copper is exclusively stored in rhogocytes (see Sect. 3) and exists in two forms, one bound to metallothioneins (Cu-MT-bound) and another in granular form (granular Cu). If the Cu-MT-bound remains always stable in relation to external metal supplies, the granular Cu is able to fluctuate rapidly in response to any outside copper influx. The presence of Cu-MT in rhogocytes can therefore be associated with the physiological regulation of copper, whereas granular precipitation is more often related to excessive physiological metal exposure. These data indicate that the copper form inside the rhogocytes (i.e., Cu-MT-bound or granular Cu) can serve as a specific biomarker to disclose copper toxicity to land snails.

The extent to which terrestrial pulmonate MTs can be applied as specific biomarkers for metal pollution under field conditions is still unknown (Dabrio et al. 2002). Both endogenous (e.g., age, body weight, size, seasonal cycle of activity, physiological changes related to juvenile–adult transition) and exogenous factors (e.g., photoperiod, vapor pressure, temperature, mean rainfall level, soil physicochemical properties, location) are known to influence metal body burden (e.g., Williamson 1980; Ireland 1984; Menta and Parisi 2001; Mourier et al. 2011). Nutritional or developmental factors could also impede MT synthesis (Dallinger 1994). As long as the impact of these factors on induction of MTs are not entirely understood, only snails of similar size, and specimens collected from the same area during the same season should be used in biomonitoring studies of copper pollution in terrestrial ecosystems. Dallinger et al. (2001) suggested that a possible approach to optimize MTs use as biomarkers of pollution is to correlate MT concentration in land snails with sublethal parameters of toxicity such as growth or fertility.

5.4 Assessment of Oxidative Stress (OS)

Heavy metal toxicity results in heightened production of intracellular reactive oxygen species (ROS) in biological systems (Valko et al. 2005). Oxidative stress (OS) defines the potential of ROS to damage cellular components such as biomembranes, proteins, DNA, and RNA (Radwan et al. 2010). Chemically, this phenomenon interferes with cellular redox balance (Jones 2006) by increasing the production of oxidizing species (e.g., xanthine oxidase, NADPH oxidases, cytochromes P450) and by decreasing antioxidant levels (e.g., catalase, CAT; glutathione reductase, GR; glutathione S-transferase, GST; glutathione peroxidase, GPx; total glutathione, GSH; superoxide dismutase, SOD) (Schafer and Buettner 2001). The quantitative measurement of intracellular ROS production and the balance between pro- and antioxidant factors in biological systems may therefore be useful as biomarkers for assessing the oxidative damage induced by heavy metal exposure. Other OS biomarkers are associated with the capacity of cellular antioxidants to neutralize specific ROS such as peroxyl radicals (ROO˙) and hydroxyl radicals (HO˙) and are also associated with the loss of DNA integrity (Moore et al. 2004; Regoli 2000; Regoli et al. 2004). Regoli et al. (2006) used *Helix aspersa* as a sentinel organism for active biomonitoring of urban airborne pollutants (see Sect. 2.1). The average concentration of trace copper in the digestive glands of these snails was directly related to the mean level of antioxidant enzymes (CAT, GR) and total oxyradical scavenging capacity (TOSC) toward ROO˙and HO˙(Table 4). The land snail, *Theba pisana*, exposed to copper via epithelial contact, showed a dose-dependent increase of GSH activity. In contrast, as shown in Table 4, GST and GPx levels followed an opposite trend, decreasing as copper concentration increased (Radwan et al. 2010). Similar responses were commonly encountered in other mollusk species when they were exposed to Cu (Doyotte et al. 1997; Alves de Almeida et al. 2004; Chandran et al. 2005). Changes in antioxidant levels can therefore be used as biomarkers of Cu-mediated prooxidant challenge in land snails.

In both lab and field studies, Itziou et al. (2011) reported an increase in ROS production, in protein-bound carbonyl (PCC) frequency and DNA strand breaks (SB) in the digestive cells and haemocytes of the snail *Eobania vermiculata* after 25 days of Cu exposure. Because the formation of protein carbonyl groups is an irreversible process (Sheehan 2006), assessing oxidative protein modifications is regarded to be a more sensitive OS biomarker than antioxidant enzyme activity (Prevodnik et al. 2007). Although measuring SB in DNA is a sensitive indicator of genotoxicity (Itziou et al. 2011), SB has limitations as a biomarker of Cu exposure in land snails, because it results from a wide range of agents and mechanisms (Mitchelmore and Chipman 1998).

Most land snails are subject to annual cycles of activity that intersperse long periods of dormancy (hibernation/estivation) with periods of activity (Nowakowska et al. 2009; Ferreira-Cravo et al. 2010). Such variations are expected to induce a marked seasonal OS. However, because of their elevated sensitivity, OS biomarkers can be used as a proper tool to detect early signs of biological disturbance that is induced by copper pollution.

6 Future Perspectives for Research on Land Snails (*Pulmonata*) as Bioindicators of Copper Pollution in Terrestrial Ecosystems

The studies reviewed in this article attest to the ability of land snails to efficiently absorb, tolerate, and regulate high amounts of copper ions inside their bodies. Depending on the concentration, copper can function at the cellular level either as a vital or toxic element for gastropod metabolism. However, physiological particularities (e.g., species, age, duration and exposure, and path of contamination) may be more deeply involved in regulating terrestrial gastropod copper concentrations than we actually know. Therefore, to fully evaluate land snail validity, reliability, and utility as bioindicators of copper pollution in terrestrial ecosystems further studies are required to provide a more profound understanding of how copper is regulated and detoxified in terrestrial gastropods. In this regard, we propose the following:

1. Because little information exists concerning the saturation limit of copper in binding to Cu-MT (Dallinger et al. 2000), *future studies should be performed to evaluate Cu-MT specificity and sensitivity for quantifying tissue levels over which copper might exert toxic effects on terrestrial snails.* In rhogocytes copper occurs in two forms, the Cu-MT-bound form, which is independent of external copper uptake, and the granular Cu form, which can fluctuate in response to over physiological copper exposure (Dallinger et al. 2005). Copper uptake, intracellular trafficking, and detoxification in terrestrial pulmonates appear to be restricted only to rhogocytes (Dallinger et al. 2005), but the cellular background of this process has not yet been discovered. There is, however, evidence to suggest that endocytosis, exocytosis, diapedesis, or carrier proteins may contribute to copper regulation in terrestrial pulmonates (Brown 1967; Marjorette et al. 1998; Marigomez et al. 2002; Riggio et al. 2002).

2. *Investigations are needed to address the existence of the genetic background for copper tolerance in land snails and also the specific effects that copper has on allele frequencies at particular loci.* Genetic mechanisms for copper tolerance have been reported in several species of invertebrates, such as earthworms (Fisker et al. 2011) and crustaceans (Lopes et al. 2005). Similar mechanisms are also known to exist in aquatic gastropods. For example, Kurelec et al. (1995) reported an increased activity of multixenobiotic resistance mechanism (MXRM) to the marine snail *Monodonta turbinata* living in polluted areas. Allozyme loci (i.e., allele genes) may be related to the level of pollution in the environment (D'Surney et al. 2001), but heavy metal exposure did not seem to account for new allozyme variants in terrestrial snails (Mulvey et al. 1996; Jordaens et al. 2006).

3. *Research concerning the use of pulmonates for monitoring copper pollution in terrestrial ecosystems should be extended to other land snail species* that are routinely employed as a sentinel organism for monitoring environmental pollution (e.g., *Cepaea* species, *Achatina fulica*). *Helix aspersa* is the only land snail species for which reference values, concerning the level to which copper accumulates in the snail body, have been proposed: 72 ppm in the snail foot and 62 ppm in the snail viscera (Pihan and de Vaufleury 2000). Palacios et al. (2011) showed that although they are different, the Cu-MTs belonging to *H. pomatia* and *H. aspersa* share a common ancestry, and suggested that their evolution over time was achieved exclusively by modulation of non-cysteine amino acid positions. Gomot and Pihan (1997) found that the capacity to accumulate copper varies among different species of land snails: *H. aspersa maxima > H. aspersa maxima ≥ H. pomatia*. Depending on the objective of the study, the most adapted species could be either autochthonous species, really living in the contaminated area (passive biomonitoring), and/or a sentinel species of known biological past ("snail watch" as proposed by De Vaufleury et al. 2012), that are not always

found in this area but can be used to reveal the current availability of copper in the environment.

4. In addition to commonly used biomarkers (see Sects. 5.1–5.4), *several other biomarkers deserve to be further investigated for assessing the impact of copper pollution on land snails*. For example, it was found that copper exposure can decrease acetylcholinesterase and ATP-ase activity in the digestive glands of land snails (De Souza Dahm et al. 2006; Itziou and Dimitriadis 2011). Heat shock protein HSP 70 may also serve as a potential bioindicator of copper exposure as Köhler et al. (1996) reported that slugs (i.e., *Deroceras reticulatum*) exposed to Zn, Pb, or Cd exhibited an elevated expression of HSP70 biosynthesis.

Such fundamental work as described above must be accomplished prior to standardizing a targeted screening procedure to use land snails in an environmental quality grid as bioindicator organisms of copper hazard on terrestrial ecosystems.

7 Summary

Land snails (*Pulmonata*) are one of the most often employed bioindicator species in environmental analysis and are considered to be a valuable invertebrate model for use in ecotoxicological and toxicological studies. Although copper plays a double role in all live beings (i.e., acting as a vital or toxic element, depending on concentration, chemical, and physical form), its environmental occurrence at high levels represents a potentially serious hazard for human health. Cu also poses risks to land snails, wherein this element is deeply involved in metabolic activities, particularly as a component of the chromoprotein hemocyanin, which is essential to respiration. Cu can enter into the land snail body directly through the tegument or the lungs, as well as indirectly via dietary uptake. The level to which this metal accumulates in snail tissues depends on the age, species, body part, metabolic status, and chemical compound involved. Copper uptake in terrestrial gastropods is homeostatically regulated through efficient metabolic mechanisms that rely on cellular compartmentalization and intratissular sequestration by complexation to specific metallothioneins (Cu-MT). Therefore, land snails are not suitable bioindicators for use in measuring long-term Cu exposure.

However, a wide range of biomarkers can be used to assess the effect of short-term copper exposure on land snails (*Pulmonata*). The induction of metallothioneins and the precipitation of copper inside rhogocytes in response to outside copper influx provide the most targeted and efficient way to quantify copper hazard on land snails. Lysosomal stability in gastropod haemocytes serves as an early warning system for detecting copper exposure in terrestrial ecosystems. Among various biomarkers of oxidative stress resulting from copper exposure, the quantification of oxidative protein modifications is the most sensitive tool for monitoring copper pollution. In addition, ultrastructural, histopathological, and genotoxic alterations that are induced by copper exposure reveal qualitative data concerning how land snails are physiologically able to regulate Cu in their bodies.

Despite regular employment in environmental biomonitoring, using land snail enzymes and metallothioneins as an effect criterion to indicate copper pollution has limitations. The limits originate from the variability in metabolic activity that exists for wild land snails. To overcome this disadvantage, researchers must follow a step-by step approach to distinguish the most appropriate biomarkers for use in biomonitoring of copper pollution. What biomarkers are selected will depend upon the aims of the research and the experimental set-up. In addition, if the stage of the land snails used is well described (e.g., origin, age, size, season of sampling, activity status) any drawbacks that arise can be easily reduced, allowing for employment of land snails as reliable bioindicators of short-term copper exposure in terrestrial ecosystems.

Acknowledgments All authors have equal rights and have contributed evenly to this paper. The present work was funded by project "Post-doctoral School Of Agriculture And Veterinary Medicine Posdru/89/1.5/S/62371", co-financed by the European Social Fund through the Sectorial Operational Programme for the Human Resources Development 2007–2013. We are grateful for the anonymous reviewers' constructive comments and recommendations, which greatly helped to increase the quality of the present review. In addition, the first author (DVN) wishes to thank D.C. for continuous support throughout his scientific career.

References

Aboho SY, Anhwange BA, Ber GA (2009) Screening of *Achatina achatina* and *Pila ovata* for trace metals in Makurdi metropolis. Pak J Nutr 8(8):1170–1171

Adeyeye EI (1996) Waste yield, proximate and mineral composition of three different types of land snails found in Nigeria. Int J Food Sci Nutr 47(2):111–116

Ahearn G, Zhuang Z, Duerr J, Pennington V (1994) Role of invertebrate electrogenic 2Na+/1H+ antiporter in monovalent and divalent cation transport. J Exp Biol 196:319–335

Albrecht U, Keller H, Gebauer W, Markl J (2001) Rhogocytes (pore cells) as the site of hemocyanin biosynthesis in the marine gastropod *Haliotis tuberculata*. Cell Tissue Res 304(3):455–462

Almendros A, Porcel D (1992) Phosphatase activity in the hepatopancreas of *Helix aspersa*. Comp Biochem Physiol A 103(30):455–460

Alves de Almeida E, Miyamoto S, Celso Dias Bainy A, Gennari de Medeiros MH, Di Mascio P (2004) Protective effect of phospholipid hydroperoxide glutathione peroxidase (PHGPx) against lipid peroxidation in mussels *Perna perna* exposed to different metals. Mar Pollut Bull 49:386–392

Amaral AFS, Anselmo H, Tristão Da Cunha RMPT, Rodrigues AS (2004) The connective tissue index of *Helix aspersa* as a metal biomarker. Biometals 17(6):625–629

Amusan AAS, Anyaele OO, Lasisi AA (2002) Effects of copper and lead on growth, feeding and mortality of terrestrial gastropod *Limicolaria flammea* (Muller, 1774). Afr J Biomed Res 5(1–2):47–50

Arndt U, Nobel W, Schweizer B (1987) Bioindikatoren: Möglichkeiten. Grenzen und neue Erkenntnisse. Ulmer Verlag, Stuttgart, pp 5–14

ATSDR (2002) Toxicological profile for copper (draft for public comment). Atlanta, GA, US Department of Health and Human Services, Public Health Service, Agency for Toxic Substances and Disease Registry (Subcontract No. ATSDR-205-1999-00024)

Babafola OO, Akinsoyinu AO (2009) Proximate composition and mineral profile of snail meat from different breeds of land snail in Nigeria. Pak J Nutr 8(12):1842–1844

Barceloux DG (1999) Copper. Clin Toxicol 37(2):217–230

Baur A, Baur B, Fröberg L (1994) Herbivory on calcicolous lichens: differential food preferences and growth rates in two co-existing land snails. Oecologia 98:313–319

Beeby A, Richmond L (2003) Do the soft tissues of *Helix aspersa* serve as a quantitative sentinel of predicted free lead concentrations in soil? Appl Soil Ecol 22(2):159–165

Beeby A, Richmond L (2002) Evaluating *Helix aspersa* as a sentinel for mapping metal pollution. Ecol Indic 1(4):261–270

Beeby A, Richmond L (1987) Adaptation by an urban population of the snail *Helix aspersa* to a diet contaminated with lead. Environ Pollut 46(1):73–82

Beeby AN (1985) The role of *Helix aspersa* as a major herbivore in the transfer of lead through a polluted ecosystem. J Appl Ecol 22:267–275

Berger B, Dallinger R, Hunziker PE, Gering E (1997) Primary structure of a copper-binding metallothionein from mantle tissue of the terrestrial gastropod *Helix pomatia* L. Biochem J 311(1):951–957

Berger B, Dallinger R (1993) Terrestrial snails as quantitative indicators of environmental metal pollution. Environ Monit Assess 25(1):65–84

Berger B, Dallinger R (1989) Accumulation of cadmium and copper by the terrestrial snail *Arianta arbustorum* L.: kinetics and budgets. Oecologia 79(1):60–65

Bertram M, Graedel TE, Rechberger H, Spatari S (2002) The contemporary European copper cycle: waste management subsystem. Ecolog Econ 42(1–2):43–57

Bigliardi E, Bertani PL, Bracchi PG (1988/1989) Contenuto di cadmio e piombo in chiocciole opercolate (*Helix pomatia* L.) raccolte nella Val Gesso (CN). Ann Fac Med Vet Parma VIII/IX:107–118

Blott EJ, Griffiths GM (2002) Secretory lysosomes. Nat Rev Mol Cell Biol 3:122–131

Bonnelly de Calventi I (1965) Copper poisoning in the snail *Helix pomatia* and its effect on mucous secretion. Ann N Y Acad Sci 118:1015–1020

Bremner I (1998) Manifestations of copper excess. Am J Clin Nutr 67(5):1069S–1073S

Brewer GJ (2007) Iron and copper toxicity in diseases of aging, particularly atherosclerosis and Alzheimer's disease. Exp Biol Med 232(2):323–335

Brown AC (1967) Elimination of foreign particles by the snail, *Helix aspersa*. Nature 213(5081):1154–1155

Brown VM, Shaw TL, Shurben DG (1974) Aspects of water quality and the toxicity of copper to rainbow trout. Water Res 8(10):797–803

Brun LA, Maillet J, Hinsinger P, Pépin M (2001) Evaluation of copper availability to plants in copper-contaminated vineyard soils. Environ Pollut 111(2):293–302

Cajaraville MP, Abascal I, Etxeberria M, Marigomez I (1995) Lysosomes as cellular markers of environmental pollution: time- and dose-dependent responses of the digestive lysosomal system of mussels after petroleum hydrocarbon exposure. Environ Toxicol Water Qual 10(1):1–8

Caldentey R, Mondschein S (2003) Policy model for pollution control in the copper industry, including a model for the sulfuric acid market. Oper Res 51(1):1–16

Carter MA, Jeffery RCV, Williamson P (1979) Food overlap in co-existing populations of the land snails *Cepaea nemoralis* (L.) and *Cepaea hortensis* (Müll). Biol J Linn Soc 11(2):169–176

CEPA (2002) Connecticut Environmental Policy Act (PA 02–121)

Chabicovsky M, Niederstaetter H, Thaler R, Hödl E, Parson W, Rossmanith W, Dallinger R (2003) Localisation and quantification of Cd and Cu-specific metallothionein isoform mRNA in cells and organs of the terrestrial gastropod *Helix pomatia*. Toxicol Appl Pharmacol 190(1):25–36

Chandran R, Sivakumar AA, Mohandass S, Aruchami M (2005) Effect of cadmium and zinc on antioxidant enzyme activity in the gastropod, *Achatina fulica*. Comp Biochem Phys C 140:422–426

Cherian MG, Goyer RA (1978) Role of metallothioneins in disease. Ann Clin Lab Sci 8(2):91–94

Chevalier L, Desbuquois C, Le Lannic J, Charrier M (2001) Les Poacées dans le régime alimentaire de l'escargot *Helix aspersa* Müller (Gastropoda, Pulmonata). C R Acad Sci, Ser III 324(11):979–987

Clayes E, Demeyer D (1986) Nutritional quality of snails (*Helix aspersa*). In: 32nd European meating of meat research workers, proceedings II, Ghent, pp 415–418

Coeurdassier M, Scheifler R, De Vaufleury A, Crini N, Saccomani C, Du Mont SL, Badot PM (2007) Earthworms influence metal transfer from soil to snails. Appl Soil Ecol 35(2):302–310

Coeurdassier M, Gomot-De Vaufleury A, Lovy C, Badot PM (2002) Is the cadmium uptake from soil important in bioaccumulation and toxic effects for snails? Ecotoxicol Environ Saf 53(3):425–431

Coughtrey PJ, Martin MH (1977) The uptake of lead, zinc, cadmium, and copper by the pulmonate mollusc, *Helix aspersa* Müller, and its relevance to the monitoring of heavy metal contamination of the environment. Oecologia 27(1):65–74

Coughtrey PJ, Martin MH (1976) The distribution of Pb, Zn, Cd and Cu within the pulmonate mollusc *Helix aspersa* Müller. Oecologia 23:315–322

Dabrio M, Rodriguez AR, Bordin G, Bebianno MJ, De Ley M, Šestáková I, Vašák M, Nordberg M (2002) Recent developments in quantification methods for metallothionein. J Inorg Biochem 88(2):123–134

Dallinger R, Chabicovsky M, Hödl E, Prem C, Hunziker P, Manzl C (2005) Copper in *Helix pomatia* (Gastropoda) is regulated by one single cell type: differently responsive metal pools in rhogocytes. Am J Physiol Regul Integr Comp Physiol 289(4):R1185–R1195

Dallinger R, Chabicovsky M, Lagg B, Schipflinger R, Weirich HG, Berger B (2004a) Isoform-specific quantification of metallothionein in the terrestrial gastropod *Helix pomatia*. II. A differential biomarker approach under laboratory and field conditions. Environ Toxicol Chem 23(4):902–910

Dallinger R, Lagg B, Egg M, Schipflinger R, Chabicovsky M (2004b) Cd accumulation and Cd-Metallothionein as a biomarker in *Cepaea hortensis* (Helicidae, Pulmonata) from laboratory exposure and metal-polluted habitats. Ecotoxicology 13(8):757–772

Dallinger R, Berger B, Triebskorn-Köhler R, Köhler H (2001a) Soil biology and ecotoxicology. In: Barker GM (ed) The biology of terrestrial molluscs. CABI Publishing, Wallingford, pp 489–525

Dallinger R, Wang Y, Berger B, Mackay E, Kägi JHR (2001b) Spectroscopic characterization of metallothionein from the terrestrial snail, *Helix pomatia*. Eur J Biochem 268(15):4126–4133

Dallinger R, Berger B, Gruber C, Stürzenbaum S (2000) Metallothioneins in terrestrial invertebrates: structural aspects, biological significance, and implications for their use as biomarkers. Cell Mol Biol 46(2):331–346

Dallinger R, Berger B, Hunziker PE, Kägi JHR (1997) Metallothionein in snail Cd and Cu metabolism. Nature 388(6639):237–238

Dallinger R (1996) Metallothionein research in terrestrial invertebrates: synopsis and perspectives. Compar Biochem Physiol Part C 113(3):125–133

Dallinger R (1994) Invertebrate organisms as biological indicators of heavy-metal pollution. Appl Biochem Biotechnol 48(1):27–31

Dallinger R, Rainbow PS (1993) Ecotoxicology of metals in invertebrates. Lewis Publishers, Boca Raton, FL, pp 245–289

Dallinger R, Janssen HH, Bauer-Hilty A, Berger B (1989) Characterization of an inducible cadmium-binding protein from hepatopancreas of metal-exposed slugs (*Arionidae*, Mollusca). Compar Biochem Physiol Part C 92(2):355–360

Dallinger R, Wieser W (1984a) Patterns of accumulation, distribution and liberation of Zn, Cu, Cd and Pb in different organs of the land snail *Helix pomatia* L. Compar Biochem Physiol Part C 79(1):117–124

Dallinger R, Wieser W (1984b) Molecular fractionation of Zn, Cu, Cd and Pb in the midgut gland of *Helix pomatia* L. Compar Biochem Physiol Part C 79(1):125–129

Dale VH, Beyeler SC (2001) Challenges in the development and use of ecological indicators. Ecol Indic 1(1):3–10

Dan N, Bailey SER (1982) Growth, mortality and feeding rates of the snail *Helix aspersa* at different population densities in the laboratory and the depresion of activity of Helicid snails by other individuals or their mucus. J Molluscan Stud 48(3):257–265

Davidson CI, Phalen RF, Solomon PA (2005) Airborne particulate matter and human health: a review. Aerosol Sci Technol 39(8):737–749

Davies DJA, Bennett BG (1985) Exposure of man to environmental copper: an exposure to contaminant assessment. Sci Total Environ 46(1–4):215–228

Davis RD, Beckett PHT (1978) Upper critical levels of toxic elements in plants II. Critical levels of copper in young barley, wheat, rape, lettuce and ryegrass, and of nickel and zinc in young barley and ryegrass. New Phytol 80(1):23–32

De Grisse A, Defloor J, Vercauteren F (1996) The influence of the addition of soils to food on the growth of the edible snail Helix aspersa maxima. Meded Fac Landbouwkd Toegep Biol Wet Univ Gent 61(1):129–139

De Souza Dahm KC, Rückert C, Tonial EM, Bonan CD (2006) In vitro exposure of heavy metals on nucleotidase and cholinesterase activities from the digestive gland of Helix aspersa. Compar Biochem Physiol Part C 143(3):316–320

De Vaufleury, Gimbert F, Pauget B, Fritsch B, Scheifler B, Coeurdassier M (2012) Les escargots bio-indicateurs de la qualite des sols—Snail watch: analyse en laboratoire ou in situ de la biodisponibilité des contaminants. http://hal.archives-ouvertes.fr/hal-00670360. Accessed 24 Aug 2012

Deng H, Ye ZH, Wong MH (2004) Accumulation of lead, zinc, copper and cadmium by 12 wetland plant species thriving in metal-contaminated sites in China. Environ Pollut 132(1):29–40

Desbuquois C, Madec L (1998) Within-clutch egg cannibalism variability in hatchlings of the land snail Helix aspersa (Pulmonata: Stylommatophora): influence of two proximate factors. Malacologia 39(1–2):167–173

Desbuquois C, Daguzan J (1995) The influence of ingestive conditioning on food choices in the land snail Helix aspersa müller (Gastropoda: Pulmonata: Stylommatophora). J Molluscan Stud 61(3):353–360

Dimitriadis VK (2001) Structure and function of the digestive system in Stylommatophora. In: Barker GM (ed) The biology of terrestrial molluscs. CABI Publishing, Wallingford, pp 237–257

Donval A, Plana S (1997) Alterations des lysosomes de la glande digestive chez bivalves de la Rade de Brest. Ann Inst Oceanogr (Paris) 73(1):69–76

Doyotte A, Cossu C, Jacquin MC, Babut M, Vasseur P (1997) Antioxidant enzymes, glutathione and lipid peroxidation as relevant biomarkers of experimental or field exposure in the gills and the digestive gland of the freshwater bivalve Unio tumidus. Aquat Toxicol 39:93–110

Druart C, Millet M, Raeppel C, Scheifler R, Delhomme O, De Vaufleury A (2011) Snails as indicators of pesticide drift, deposit, transfer and effects in the vineyard. Sci Total Environ 409(20):4280–4288

Druart C, Scheifler R, De Vaufleury A (2010) Towards the development of an embryotoxicity bioassay with terrestrial snails: screening approach for cadmium and pesticides. J Hazard Mater 184(1–3):26–33

D'Surney SJ, Shugart LR, Theodorakis W (2001) Genetic markers and genotyping methodologies: an overview. Ecotoxicology 10:201–204

Dupont-Nivet M, Coste V, Coinon P, Bonnet J, Blanc J (2000) Rearing density effect on the production performance of the edible snail Helix aspersa Müller in indoor rearing. Ann Zootech 49(5):447–456

Eeva T, Rainio K, Suominen O (2010) Effects of pollution on land snail abundance, size and diversity as resources for pied flycatcher, Ficedula hypoleuca. Sci Total Environ 408(19):4165–4169

Egonmwan RI (2008) Effects of dietary calcium on growth and oviposition of the African land snail Limicolaria flammea (Pulmonata: Achatinidae). Rev Biol Trop 56(1):333–343

El-Gendy KS, Radwan MA, Gad AF (2011) Feeding and growth responses of the snail Theba pisana to dietary metal exposure. Arch Environ Contam Toxicol 60(2):272–280

El-Gendy KS, Radwan MA, Gad AF (2009) In vivo evaluation of oxidative stress biomarkers in the land snail, Theba pisana exposed to copper-based pesticides. Chemosphere 77(3):339–344

Elmslie LJ (1998) Humic acid a growth factor for Helix aspersa Müller (Gastropoda: Pulmonata). J Molluscan Stud 64(3):400–401

Extoxnet (1996) Anon. Pesticide information profiles. ExtensionToxicology Network. Copper sulfate. http://extoxnet.orst.edu/pips/coppersu.htm. Accessed 4 Sept 2011

Fagbuaro O, Oso JA, Edward JB, Ogunleye RF (2006) Nutritional status of four species of giant land snails in Nigeria. J Zhejiang Univ Sci B 7(9):686–689

Fisker KV, Sørensen JG, Damgaard C, Pedersen KL, Holmstrup M (2011) Genetic adaptation of earthworms to copper pollution: is adaptation associated with fitness costs in *Dendrobaena octaedra*? Ecotoxicology 20(3):563–573

Finnie AA (2006) Improved estimates of environmental copper release rates from antifouling products. Biofouling 22(5):279–291

Ferreira-Cravo M, Welker Alexis F, Hermes-Lima M (2010) The connection between oxidative stress and estivation in gastropods and anurans. Prog Mol Subcell Biol 49:47–61

Forray FL (2002) Geochemistry of the environment in the areas of mining works from Aries Valley (Apuseni mountains, Romania). PhD thesis, Babes-Bolyai University

Fritsch C, Coeurdassier M, Gimbert F, Crini N, Scheifler R, De Vaufleury A (2011) Investigations of responses to metal pollution in land snail populations (*Cantareus aspersus* and *Cepaea nemoralis*) from a smelter-impacted area. Ecotoxicology 20(4):739–759

Fritsch C, Scheifler R, Beaugelin-Seiller K, Hubert P, Cœurdassier M, de Vaufleury A, Badot P-M (2008) Biotic interactions modify the transfer of cesium-137 in a soil-earthworm-plant-snail food web. Environ Toxicol Chem 27(8):1698–1707

Gadd GM (1992) Metals and microorganisms: a problem of definition. FEMS Microbiol Lett 79(1–3):197–204

García A, Perea JM, Mayoral A, Acero R, Martos J, Gómez G, Peña F (2006) Laboratory rearing conditions for improved growth of juvenile *Helix aspersa* Müller snails. Lab Anim 40(3):309–316

Gbaruko BC, Friday OU (2007) Bioaccumulation of heavy metals in some fauna and flora. Int J Environ Sci Technol 4(2):197–202

George SG, Pirie BJS, Cheyne AR, Coombs TL, Grant PT (1978) Detoxification of metals by marine bivalves: an ultrastructural study of the compartimentation of copper and zinc in the oyster *Ostrea edulis*. Mar Biol 45:147–156

Georgopoulos PG, Roy A, Yonone-Lioy MJ, Opiekun RE, Lioy PJ (2001) Environmental copper: its dynamics and human exposure issues. J Toxicol Environ Health, Part B 4(4):341–394

Gheoca V, Gheoca D (2005) The accumulation of heavy metals in the tissues of Helix pomatia from locations with industrial and town pollution. Transylvanian Rev Syst Ecol Res 2:67–74

Giamberini L, Pihan LC (1997) Lysosomal changes in the hemocytes of the freshwater mussel *Dreissena polymorpha* experimentally exposed to lead and zinc. Dis Aquat Org 28:221–227

Gil F, Pla A (2001) Biomarkers as biological indicators of xenobiotic exposure. J Appl Toxicol 21(4):245–255

Gimbert F, Vijver MG, Coeurdassier M, Scheifler R, Peijnenburg WJGM, Badot PM, De Vaufleury A (2008) How subcellular partitioning can help to understand heavy metal accumulation and elimination kinetics in snails. Environ Toxicol Chem 27(6):1284–1292

Gimbert F, de Vaufleury A, Douay F, Scheifler R, Coeurdassier M, Badot PM (2006) Modelling chronic exposure to contaminated soil: a toxicokinetic approach with the terrestrial snail *Helix aspersa*. Environ Int 32:866–875

Gogoasa I, Gergen I, Oprea G, Bordean D, Alda L, Moigradean D, Rada M, Bragea M (2011) Preliminary research concerning the distribution of copper in the soil and vegetables in historical anthropic pollution (Caras-Severin County, Romania). J Agroaliment Process Technol 17:371–374

Gomot de Vaufleury A, Coeurdassier M, Pandard P, Scheifler R, Lovy C, Crini N, Badot PM (2006) How terrestrial snails can be used in risk assessment of soils. Environ Toxicol Chem 25(3):797–806

Gomot de Vaufleury A, Kerhoas I (2000) Effects of cadmium on the reproductive system of the land snail *Helix aspersa*. Bull Environ Contam Toxicol 64:434–442

Gomot de Vaufleury A, Pihan F (2000) Growing snails used as sentinels to evaluate terrestrial environment contamination by trace elements. Chemosphere 40(3):275–284

Gomot de Vaufleury A (2000) Standardized growth toxicity testing (Cu, Zn, Pb, and Pentachlorophenol) with *Helix aspersa*. Ecotoxicol Environ Saf 46(1):41–50

Gomot A (1998) Biochemical composition of *Helix* snails: influence of genetic and physiological factors. J Molluscan Stud 64(2):173–181

Gomot A (1997) Dose-dependent efects of cadmium on the growth of snails in toxicity bioassays. Arch Environ Contam Toxicol 33(2):209–216

Gomot A, Pihan F (1997) Comparison of the bioaccumulation capacities of copper and zinc in two snail subspecies (*Helix*). Ecotoxicol Environ Saf 38:85–94

Gomot A, Gomot L, Boukraa S, Bruckert S (1989) Influence of soil on the growth of the land snail *Helix aspersa*. An experimental study of the absorption route for the stimulating factors. J Molluscan Stud 55(1):1–7

Gomot L, Deray A (1987) Les escargots. La Recherche 186:302–311

Gray JB, Kralka RA, Samuel WM (1985) Rearing of eight species of terrestrial gastropods (order Stylommatophora) under laboratory conditions. Can J Zool 63(1):2474–2476

Graveland J, Van Der Wal R, Van Balen JH, Van Noordwijk AJ (1994) Poor reproduction in forest passerines from decline of snail abundance on acidified soils. Nature 368:446–448

Greville RW, Morgan AJ (1990) The influence of size on the accumulated amounts of metals (Cu, Pb, Cd, Zn and Ca) in six species of slug sampled from a contaminated woodland site. J Molluscan Stud 56(3):355–362

Guecheva T, Henriques JAP, Erdtmann B (2001) Genotoxic effects of copper sulphate in freshwater planarian in vivo, studied with the single-cell gel test (comet assay). Mutat Res Gen Tox Environ 497(1–2):19–27

Hanley ME, Fenner M, Edwards PJ (1995) An experimental field study of the effects of mollusk grazing on seedling recruitment and survival in grassland. J Ecol 83:621–627

Harrison E, McBride MB, Bouldin DR (1999) Land application of sewage sludges: an appraisal of the US regulations. Int J Environ Pollut 11(1):1–43

Haszprunar G (1996) The molluscan rhogocyte (pore-cell, Blasenzelle, cellule nucale), and its significance for ideas on nephridial evolution. J Molluscan Stud 62(2):185–211

Hellawell JM (1986) Biological indicators of freshwater pollution and environmental management. Elsevier, London, pp 5–40

Heller J (2001) Life history strategies. In: Barker GM (ed) The biology of terrestrial molluscs. CAB International, Wallingford, pp 413–445

Hodasi JKM (1979) Life-history studies of *Achatina* (*Achatina*) *Achatina* (Linné). J Molluscan Stud 45(3):328–339

Hong S, Candelone JP, Patterson CC, Boutron CF (1996a) History of ancient copper smelting pollution during Roman and Medieval times recorded in Greenland Ice. Science 272(5259):246–249

Hong S, Candelone JP, Patterson CC, Boutron CF (1996b) A reconstruction of changes in copper production and copper emissions to the atmosphere during the past 7000 years. Sci Total Environ 188(2–3):183–193

Hopkin SP (1993a) Deficiency and excess of copper in terrestrial isopods. In: Dallinger R, Rainbow P (eds) Ecotoxicology of metals in invertebrates. Lewis Publishing, Boca Raton, pp 359–382

Hopkin SP (1993b) Ecological implications of 95% protectionlevels' for metals in soil. Oikos 66:137–141

Howard B, Mitchell PCH, Ritchie A, Simkiss K, Taylor M (1981) The composition of intracellular granules from the metal accumulating cells of the snail *Helix aspersa*. Biochem J 194(2):507–511

Höckner M, Stefanon K, de Vaufleury A, Monteiro F, Pérez-Rafael S, Palacios Ò, Capdevila M, Atrian S, Dallinger R (2011) Physiological relevance and contribution to metal balance of specific and non-specific metallothionein isoforms in the garden snail, *Cantareus aspersus*. Biometals 24(6):1079–1092

Hulmann H, Kraft U (2003) Kupfer und Zink—Die bedeutung der abschwemmungen von metalldächern. GWF-Wasser/Abwasser 144(2):127–133

Hulskotte JHJ, Denier van der Gon HAC, Visschedijk AJH, Schaap M (2007) Brake wear from vehicles as an important source of diffuse copper pollution. Water Sci Technol 56(1):223–231

Iglesias J, Castillejo J (1999) Field observations on feeding of the land snail *Helix aspersa* Müller. J Molluscan Stud 65(4):411–423

Ireland MP, Marigomez I (1992) The influence of dietary calcium on the tissue distribution of Cu, Zn, Mg & P and histological changes in the digestive gland cells of the snail *Achatina fulica* Bowdich. J Molluscan Stud 58(2):157–168

Ireland MP (1984) Seasonal changes in zinc, manganese, magnesium, copper and calcium content in the digestive gland of the slug *Arion ater* L. Comp Biochem Physiol A 78(4):855–858

Ireland MP (1979) Distribution of essential and toxic metals in the terrestrial slug *Arion ater* (L). Environ Pollut 20(4):271–278

Ismert M, Oster T, Bagrel D (2002) Effects of atmospheric exposure to naphthalene on xenobiotic-metabolizing enzymes in the snail *Helix aspersa*. Chemosphere 46(2):273–280

Itziou A, Dimitriadis V (2011) Introduction of the land snail *Eobania vermiculata* as a bioindicator organism of terrestrial pollution using a battery of biomarkers. Sci Total Environ 409(6):1181–1192

Itziou A, Kaloyianni M, Dimitriadis V (2011) In vivo and in vitro effects of metals in reactive oxygen species production, protein carbonylation, and DNA damage in land snails *Eobania vermiculata*. Arch Environ Contam Toxicol 60(4):697–707

Janssen HH (1985) Some histophysiological findings on the midgut gland of the common garden snail, findings on the midgut gland of the common garden snail, *Arion rufus* (L.) (Syn. *A. ater rufus* (L.), *A. empiricorum* Férussac), *Gastropoda: Stylommatophora*. Zoolog Anz 215(1–2):33–51

Jatto OE, Asia IO, Medjor WE (2010) Proximate and mineral composition of different species of snail shell. PJTS 11(1):416–419

Jeffery J, Uren NC (1983) Copper and zinc species in the soil solution and the effects of soil pH. Aust J Soil Res 21(4):479–488

Jess S, Mark RJ (1989) The interaction of diet and substrate on the growth of *Helix aspersa* (Müller) var *Maxima*. Slug Snails World Agric 41:311–317

Jezierska B, Witeska M (2006) The metal uptake and accumulation in fish living in polluted waters. Nato Sci S SS IV Ear 69:107–114

Jokanović M, Tojagić S, Kevrešan Ž (2006) Toxic residues in controlled production of vineyard snail (*Helix pomatia*). Ann Facult Eng Hunedoara 4(3):223–226

Jones DP (2006) Redefining oxidative stress. Antioxid Redox Signal 8:1865–1879

Jones DT (1991) Biological monitoring of metal pollution in terrestrial ecosystems. PhD thesis, University of Reading

Jong-Brink M, Ankie W, Kraal G, Boer HH (1976) A light and electron microscope study on oogenesis in the freshwater pulmonate snail *Biomphalaria glabrata*. Cell Tissue Res 171(20):195–219

Jordaens K, De Wolf H, Van Houtte N, Vandecasteele B, Backeljau T (2006) Genetic variation in two land snails, *Cepaea nemoralis* and *Succinea putris* (Gastropoda, Pulmonata), from sites differing in heavy metal content. Genetica 128(1–3):227–239

Jorquera H (2009) Source apportionment of PM_{10} and $PM_{2.5\ at}$ Tocopilla, Chile (22 degrees 05' S, 70 degrees 12' W). Environ Monit Assess 153(1–4):235–251

Kabala C, Singh BR (2000) Fractionation and mobility of copper, lead, and zinc in soil profiles in the vicinity of a copper smelter. J Environ Qual 30(2):485–492

Kägi JHR, Schäffer A (1988) Biochemistry of metallothionein. Biochemistry 27(23):8509–8515

Kalyani R (1990) The effect of feeding copper sulphate to *Achatina fulica* (Pulmonata: Stylommatophora) on albumen gland polysaccharides (Copper sulphate/*Achatina fulica*/albumen gland polysaccharides). Proc Indian Natl Sci Acad B Biol Sci 56(4):335–338

Karayakar F, Cicik B, Ciftci N, Karaytug S, Erdem C, Ozcan AY (2010) Accumulation of copper in liver, gill and muscle tissues of *Anguilla anguilla* (Linnaeus, 1758). J Anim Vet Adv 9(17):2271–2274

Karlin KD, Cruse RW, Gultneh Y, Farooq A, Hayes JC, Zubieta J (1987) Dioxygen-copper reactivity. Reversible binding of O2 and CO to a phenoxo-bridged dicopper (I) complex. J Am Chem Soc 109(9):2668–2679

Klobučar GIV, Lajtner J, Erben R (1997) Lipid peroxidation and histopathological changes in the digestive gland of a freshwater snail *Planorbarius corneus* L. (Gastropoda, Pulmonata) exposed to chronic and sub-chronic concentrations of PCP. Bull Environ Contam Toxicol 58(1):128–134

Köhler HR, Rahman B, Graff S, Berkus M, Triebskorn R (1996) Expression of the stress-70 protein family (HSP70) due to heavy metal contamination in the slug, *Deroceras reticulatum*: an approach to monitor sublethal stress conditions. Chemosphere 33:1327–1340

Królak E (2003) Accumulation of Zn, Cu, Pb and Cd by dandelion (*Taraxacum officinale* web.) in environments with various degrees of metallic contamination. Pol J Environ Stud 12(6):713–721

Kuffner IB, Andersson AJ, Jokiel P, Rodgers KS, Mackenzie FT (2008) Decreased abundance of crustose coralline algae due to ocean acidification. Nat Geosci 1:114–117

Kurelec B, Lucić D, Pivčević B, Krča S (1995) Induction and reversion of multixenobiotic resistance in the marine snail *Monodonta turbinate*. Mar Biol 123(2):305–312

Kurihara Y, Suzuki T, Moriyama K (1987) Incorporation of heavy metals by the mud snail, *Cipangopaludina chinensis malleata* Reeve, in submerged paddy soil treated with composted sewage sludge. Biol Fert Soils 5(2):93–97

Landner L, Lindestrom L (1999) Copper in society and in the environment. Vasteras, Swedish Environmental Research Group (MFG) (SCDA S-721 88)

Laskowski R, Hopkin SP (1996a) Effect of Zn, Cu, Pb, and Cd on fitness in snails (*Helix aspersa*). Ecotoxicol Environ Saf 34(1):59–69

Laskowski R, Hopkin SP (1996b) Accumulation of Zn, Cu, Pb and Cd in the garden snail (*Helix aspersa*): implications for predators, 289–297. Environ Pollut 91(3):289–297

Lazaridou-Dimitriadou M, Alpoyanni E, Baka M, Brouziotis TH, Kifonidis N, Mihaloudi E, Sioula E, Vellis G (1998) Growth, mortality and fecundity in successive generations of *Helix aspersa* Müller cultured indoors and crowding effect on fast-, medium- and slow-growing snails of the same clutch. J Molluscan Stud 64(1):67–74

Lazaridou-Dimitriadou M, Kattoulas ME (1991) Energy flux in a natural population of the land snail *Eobania vermiculata* (Müller) (Gastropoda: Pulmonata: Stylommatophora) in Greece. Can J Zool 69(4):881–891

Lemos VA, Novaes GDS, De Carvalho AL, Gama EM, Santos AG (2009) Determination of copper in biological samples by flame atomic absorption spectrometry after precipitation with Me-BTAP. Environ Monit Assess 148(1–4):245–253

Lepp NW, Salmon D (1999) A field study of the ecotoxicology of copper to bryophytes. Environ Pollut 106(2):153–156

Linder MC, Hazegh-Azam M (1996) Copper biochemistry and molecular biology. Am J Clin Nutr 63(5):781–797

Linhart YB, Thompson JD (1995) Terpene-based selective herbivory by *Helix aspersa* (Mollusca) on *Thymus vulgaris* (Labiatae). Oecologia 102(1):126–132

Lippard SJ, Berg JM (1994) Principles of bioinorganic chemistry. University Science, Mill Valley, CA, pp 237–243

Livingstone DR, Chipman JK, Lowe DM, Minier C, Mitchelmore CL, Moore MN, Peters LD, Pipe RK (2000) Development of biomarkers to detect the effects of organic pollution on aquatic invertebrates: recent molecular, genotoxic, cellular and immunological studies on the common mussel (*Mytilus edulis* L.) and other mytilids. Int J Environ Pollut 13(1–6):56–91

Lopes I, Baird DJ, Ribeiro R (2005) Genetically determined resistance to lethal levels of copper by *Daphnia longispina*: association with sublethal response and multiple/coresistance. Environ Toxicol Chem 24:1414–1419

Lopes TM, Barcarolli IM, De Oliveira CB, De Souza MM, Bianchini A (2011) Mechanisms of copper accumulation in isolated mantle cells of the marine clam *Mesodesma mactroides*. Environ Toxicol Chem 30(7):1586–1592

Lowe DM, Pipe RK (1994) Contaminant induced lysosomal membrane damage in marine mussel digestive cells: an in vitro study. Aquat Toxicol 30(4):357–365

Luchtel DL, Deyrup-Olsen J (2001) Body wall: form and function. In: Barker GM (ed) The biology of terrestrial molluscs. CABI Publishing, Hamilton, pp 147–178

Luzio JP, Pryor PR, Bright NA (2007) Lysosomes: fusion and function. Nat Rev Mol Cell Biol 8:622–632

Łysak A, Mach-Paluszkiewicz Z, Ligaszewski M (2000) Production quality of edible snail *Helix aspersa maxima* in different farm management systems. Roczniki Naukowe Zootechniki 8:187–191

Machin J (1977) Role of integument in molluscs. In: Gupta BL et al (eds) Transport of ions and water in animals. Academic, New York, pp 735–762

Manzl C, Krumschnabel G, Schwarzbaum PJ, Dallinger R (2004) Acute toxicity of cadmium and copper in hepatopancreas cells from the Roman snail (*Helix pomatia*). Comp Biochem Physiol C Pharmacol Toxicol Endocrinol 138(1):45–52

Marigomez I, Soto M, Cajaraville MP, Angulo E, Giamberini L (2002) Cellular and subcellular distribution of metals in molluscs. Microsc Res Tech 56(5):358–392

Marigòmez JA, Angulo E, Moya J (1986) Copper treatment of the digestive gland of the slug, *Arion ater* L. 1. Bioassay conduction and histochemical analysis. Bull Environ Contam Toxicol 36(1):600–607

Marjorette M, Pena O, Koch KA, Thiele DJ (1998) Dynamic regulation of copper uptake and detoxification genes in *Saccharomyces cerevisiae*. Mol Cell Biol 18:2514–2523

Markert B (2007) Definitions and principles for bioindication and biomonitoring of trace metals in the environment. J Trace Elem Med Biol 21(1):77–82

Markl J (1986) Evolution and function of structurally diverse subunits in the respiratory protein hemocyanin from arthropods. Biol Bull 171(1):90–115

Marks GW (1938) The copper content and copper tolerance of some species of mollusks of the Southern California Coast. Biol Bull 75:224–237

Martin MH, Coughtrey J (1982) Use of terrestrial animals as monitors and indicators. In: Mellamby K (ed) Biological monitoring of heavy metal pollution. Applied Science, London, pp 253–271

Mason CF (1970) Snail populations, beech litter production and the role of snails in litter decomposition. Oecologia 5(3):215–239

Maynard R (2004) Key airborne pollutants—the impact on health. Sci Total Environ 334–335:9–13

McIntosh LM, Robinson WE (1999) Cadmium turnover in the hemocytes of *Mercenaria mercenaria* (L.) in relation to hemocyte turnover. Comp Biochem Physiol C Pharmacol Toxicol Endocrinol 123(1):61–66

Menezes MÂBC, Sabino CVS, Franco MB, Maia ECP, Albinati CCB (2004) Assessment of workers' contamination caused by air pollution exposure in industry using biomonitors. J Atmos Chem 49(1–3):403–414

Menta C, Parisi V (2001) Metal concentrations in *Helix pomatia*, *Helix aspersa* and *Arion rufus*: a comparative study. Environ Pollut 115(2):205–208

Mercer JFB, Llanos RM (2003) Molecular and cellular aspects of copper transport in developing mammals. J Nutr 133:1481S–1484S

Michel X, Mora P, Garrigues P, Budzinski H, Raoux C, Narbonne JF (2001) Cytochrome P450 dependent activities in mussel and fish from coastal marine environment: field studies on the French Coast of the Mediterranean Sea. Polycyclic Aromat Comp 18(3):307–324

Mighall TM, Abrahams PW, Grattan JP, Hayes D, Timberlake S, Forsyth S (2002) Geochemical evidence for atmospheric pollution derived from prehistoric copper mining at Copa Hill, Cwmystwyth, mid-Wales, UK. Sci Total Environ 292(1–2):69–80

Mikkelsen FF, Weber RE (1992) Oxygen transport and hemocyanin function in the pulmonate land snail, *Helix pomatia*: physiological and molecular implications of polyphasic oxygen-binding curves. Physiol Zool 65(6):1057–1073

Mitchelmore CL, Chipman JK (1998) DNA strand breakage in aquatic organisms and the potential value of the Comet assay in environmental monitoring. Mutat Res 399:135–147

Moore MN, Depledge MH, Readman JW, Paul Leonard DR (2004) An integrated biomarker-based strategy for ecotoxicological evaluation of risk in environmental management. Mutat Res 552(1–2):247–268

Mortazavi F, Jafari-Javid A (2009) Acute renal failure due to copper sulphate poisoning: a case report. Iran J Pediatr 19(1):75–78

Moser H, Wieser W (1979) Copper and nutrition in *Helix pomatia* (L.). Oecologia 42(2):241–251

Mourier B, Fritsch C, Dhivert E, Gimbert F, Cœurdassier M, Pauget B, de Vaufleury A, Scheifler R (2011) Chemical extractions and predicted free ion activities fail to estimate metal transfer from soil to field land snails. Chemosphere 85(6):1057–1065

Mulvey M, Newman MC, Beeby AN (1996) Genetic and conchological comparison of snails (*Helix aspersa*) differing in shell deposition of lead. J Molluscan Stud 62(2):213–223

Nanda Kumar PBA, Dushenkov V, Motto H, Raskin I (1995) Phytoextraction: the use of plants to remove heavy metals from soils. Environ Sci Technol 29(5):1232–1238

Nenițescu CD (1972) Chimie generala. Editura Didactică și Pedagogică, București, pp 89–105

Nica DV, Bura M, Gergen I, Harmanescu M, Bordean DM (2012) Bioaccumulative and conchological assessment of heavy metal transfer in a soil-plant-snail food chain. Chem Cent J 6(1):55

Nieboer E, Richardson DHS (1980) The replacement of the nondescript term 'heavy metals' by a biologically and chemically significant classification of metal ions. Environ Pollut 1(1):3–26

Nogué S, Sanz P, Munné P, Gadea E (2000) Copper contamination from domestic tap water with a descaler. Bull World Health Organ 78(4):565–566

Notten MJ, Oosthoek AJ, Rozema J, Aerts R (2006) Heavy metal pollution affects consumption and reproduction of the landsnail Cepaea nemoralis fed on naturally polluted *Urtica dioica* leaves. Ecotoxicology 15(3):295–304

Notten MJ, Oosthoek AJ, Rozema J, Aerts R (2005) Heavy metal concentrations in a soil-plant-snail food chain along a terrestrial soil pollution gradient. Environ Pollut 138(1):178–190

Nowakowska A, Świderska-Kołacz G, Rogalska J, Caputa M (2009) Antioxidants and oxidative stress in Helix pomatia snails during estivation. Comp Biochem Physiol C Pharmacol Toxicol Endocrinol 150(4):481–486

Oehlmann J, Schulte-Oehlmann U (2002) Chapter 17: Molluscs as bioindicators. In: Markert BA, Breure AM, Zechmeister HG (eds) Bioindicators and biomonitors. Elsevier Science B.V., Oxford pp 577–635

Ogiyama S, Sakamoto K, Suzuki H, Ushio S, Anzai T, Inubushi K (2005) Accumulation of zinc and copper in an arable field after animal manure application. Soil Sci Plant Nutr 51(6):801–808

Otitoloju AA, Ajikobi DO, Egonmwan RI (2009) Histopathology and bioaccumulation of heavy metals (Cu & Pb) in the Giant land snail, *Archachatina marginata* (Swainson). Open Environ Pollut Toxicol J 1:79–88

Oyedepo TA, Adeboye OO, Idowu CB (2007) Comparative study on bioaccumulation of toxic metals in molluscs and crustaceans. Bullet Sci Assoc Nigeria 28:75–78

Özogul Y, Özogul F, Olgunoglu AI (2005) Fatty acid profile and mineral content of the wild snail (*Helix pomatia*) from the region of the south of the Turkey. Eur Food Res Technol 221(3–4):547–549

Palacios Ò, Pagani A, Pérez-Rafael S, Egg M, Höckner M, Brandstätter A, Capdevila M, Atrian S, Dallinger R (2011) Shaping mechanisms of metal specificity in a family of metazoan metallothioneins: evolutionary differentiation of mollusk metallothioneins. BMC Biol 9:4

Pase L, Voskoboinik I, Greenough M, Camarkaris J (2004) Copper stimulates trafficking of a distinct pool of the Menkes copper ATPase (ATP7A) to the plasma membrane and diverts it into a rapid recycling pool. Biochem J 378(Pt 3):1031–1037

Pauget B, Gimbert F, Coeurdassier M, Scheifler R, de Vaufleury A (2011) Use of chemical methods to assess Cd and Pb bioavailability to the snail Cantareus aspersus: a first attempt taking into account soil characteristics. J Hazard Mater 192:1804–1811

Paustenbach D, Galbraith D (2006) Biomonitoring and biomarkers: exposure assessment will never be the same. Environ Health Perspect 114(8):1143–1149

Pawson P, Chase R (1984) The life-cycle and reproductive activity of Achatina fulica (Bowdich) in laboratory culture. J Molluscan Stud 50(2):85–91

Payne GG, Martens DC, Kornegay ET, Lindemann MD (1988) Availability and form of copper in three soils following eight annual applications of copper-enriched swine manure. J Environ Qual 17(4):740–746

Peña SC, Pocsidio GN (2008) Accumulation of copper by golden apple snail Pomacea canaliculata Lamarck. Philipp J Sci 137(2):153–158

Perez A, Calvo de Anta R (1992) Soil pollution in copper sulphide mining areas in Galicia (N.W. Spain). Soil Technol 5(3):271–281

Peters LD, Shaw JP, Nott M, O'Hara SCM, Livingstone DR (1999) Development of cytochrome P450 as a biomarker of organic pollution in *Mytilus* sp.: field studies in United Kingdom ('*Sea Empress*' oil spill) and the Mediterranean Sea. Biomarkers 4(6):425–441

Pietrzak U, McPhail DC (2004) Copper accumulation, distribution and fractionation in vineyard soils of Victoria, Australia. Geoderma 122(2–4):151–166

Pihan F, de Vaufleury A (2000) The snail as a target organism for the evaluation of industrial waste dump contamination and the efficiency of its remediation. Ecotoxicol Environ Saf 46(2):137–147

Pihan JC, Morhain E, Pihan F (1994) Recherche des micropolluants métalliques (cuivre et zinc) dans l'escargot en milieu naturel et en e´levage. Evaluation du facteur de contamination par la voie alimentaire (Cu, Zn, Cr, Fe, Mn). Journée Nationale Hélicicole, ITAVI, GNPE, Rennes, 1–10

Pimentel D (1971) Ecological effects of pesticides on nontarget species. Executive Office of the President's Office of Science and Technology. Washington DC, U. S. Government Printing Office

Pivarov A, Drozdowa E (2002) Dose-dependence of the excitatory effects of acetylocholine on common snail neurons after orthodromic tetanization. Neurosci Behav Physiol 32:41–43

Pizarro F, Olivares M, Uauy R, Contreras P, Rebelo A, Gidi G (1999) Acute gastrointestinal effects of graded levels of copper in drinking water. Environ Health Perspect 107(2):117–121

Porębska G, Ostrowska A (1999) Heavy metal accumulation in wild plants: implications for phytoremediation. Pollut J Environ Stud 8(6):433–442

Prevodnik A, Gardestrom J, Lilja K, Elfwing T, McDonagh B, Petrovic N et al (2007) Oxidative stress in response toxenobiotics in the blue mussel *Mytilus edulis* L.: evidence for variation along a natural salinity gradient of the Baltic Sea. Aquat Toxicol 82:63–71

Pyatt FB, Pyatt AJ, Pentreath VW (1997) Distribution of metals and accumulation of lead by different tissues in the freshwater snail *Lymnaea stagnalis* (L.). Environ Toxicol Chem 16(6):1393–1395

Rabitsch WB (1996) Metal accumulation in terrestrial pulmonates at a lead/zinc smelter site in Arnoldstein, Austria. Bull Environ Contam Toxicol 56(5):734–741

Radwan MA, El-Gendy KS, Gad AF (2010) Oxidative stress biomarkers in the digestive gland of *Theba pisana* exposed to heavy metals. Arch Environ Contam Toxicol 58(3):828–835

Ramalingam K, Indra D (2002) Copper sulphate ($CuSO_4$) toxicity on tissue phosphatases activity and carbohydrates turnover in *Achatina fulica*. J Environ Biol 23(2):181–188

Regoli F, Gorbi S, Fattorini D, Tedesco S, Notti A, Machella N, Bocchetti R, Benedetti M, Piva F (2006) Use of the land snail *Helix aspersa* as sentinel organism for monitoring ecotoxicologic effects of urban pollution: an integrated approach. Environ Health Perspect 114(1):63–69

Regoli F, Frenzilli G, Bocchetti R, Annarumma F, Scarcelli V, Fattorini D, Nigro M (2004) Time-course variation in oxyradical metabolism, DNA integrity and lysosomal stability in mussels, *Mytilus galloprovincialis*, during a field translocation experiment. Aquat Toxicol 68(2):167–178

Regoli F (2000) Total oxyradical scavenging capacity (TOSC) in polluted and translocated mussels: a predictive biomarker of oxidative stress. Aquat Toxicol 50(4):351–361

Regoli F (1998) Trace metals and antioxidant enzymes in gills and digestive gland of the mediterranean mussel *Mytilus galloprovincialis*. Arch Environ Contam Toxicol 34(1):48–63

Riggio M, Lee J, Scudiero R, Parisi E, Thiele DJ, Filosa S (2002) High affinity copper transport protein in the lizard *Podarcis sicula*: molecular cloning, functional characterization and expression in somatic tissues, follicular oocytes and eggs. Biochim Biophys Acta 1576(1–2):127–135

Robertson J (1964) Osmotic and ionic regulation. In: Wilbur K, Younge C (eds) Physiology of mollusca, vol 1. Academic, London, pp 283–311

Ryder TA, Bowen ID (1977a) The slug foot as a site of uptake copper molluscocicide. J Inv Path 30(3):381–386

Ryder TA, Bowen ID (1977b) Endocytosis and aspects of autophagy in the foot epithelium of the slug *Agriolimax reticulatus* (Müller). Cell Tissue Res 181:129–142

Rusjan D, Strlič M, Pucko D, Korošec-Koruza Z (2007) Copper accumulation regarding the soil characteristics in Sub-Mediterranean vineyards of Slovenia. Geoderma 141(1–2):110–118

Russell LK, DeHaven JI, Botts RP (1981) Toxic effects of cadmium on the Garden snail (*Helix aspersa*). Bull Environ Contam Toxicol 26:634–640

Sacks JD, Stanek LW, Luben TJ, Johns DO, Buckley BJ, Brown JS, Ross M (2011) Particulate matter–induced health effects: who is susceptible? Environ Health Perspect 119(4):446–454

Samecka-Cymerman A, Kempers A (2007) Heavy metals in aquatic macrophytes from two small rivers polluted by urban, agricultural and textile industry sewages SW Poland. Arch Environ Contam Toxicol 53(2):198–206

Sánchez de la Campa AM, De La Rosa JD, Sánchez-Rodas D, Oliveira V, Alastuey A, Querol X, Gómez-Ariza JL (2008) Arsenic speciation study of PM2.5 in an urban area near a copper smelter. Atmos Environ 42(26):6487–6495

Sánchez-Rodas D, Sánchez de la Campa AM, De la Rosa JD, Oliveira V, Gómez-Ariza JL, Querol X, Alastuey A (2007) Arsenic speciation of atmospheric particulate matter (PM_{10}) in an industrialised urban site in southwestern Spain. Chemosphere 66(8):1485–1493

Sauvé S, Dumestre A, McBride MB, Hendershot WH (1998) Derivation of soil quality criteria using predicted chemical speciation of Pb^{2+} and Cu^{2+}. Environ Toxicol Chem 17(8):1481–1489

Sauvé S, McBride MB, Norvell WA, Hendershot WH (1997) Copper solubility and speciation of in situ contaminated soils: effect of copper level, pH and organic matter. Water Air Soil Pollut 100(1–2):133–149

Schafer FQ, Buettner GR (2001) Redox environment of the cell as viewed through the redox state of the glutathione disulfide/glutathione couple. Free Radic Biol Med 30(11):1191–1212

Scheifler R, de Vaufleury A, Coeurdassier M, Crini N, Badot P-M (2006) Transfer of Cd, Cu, Ni, Pb, and Zn in a soil–plant–invertebrate food chain: a microcosm study. Environ Toxicol Chem 25(3):815–822

Scheifler RM, Brahim B, Gomot-De Vaufleury A, Carnus J, Badot PM (2003) A field method using microcosms to evaluate transfer of Cd, Cu, Ni, Pb and Zn from sewage sludge amended forest soils to *Helix aspersa* snails. Environ Pollut 122(3):343–350

Schumann K, Classen HG, Dieter HH, Konig J, Multhaup G, Rukgauer M, Summer KH, Bernhardt J, Biesalski HK (2002) Hohenheim consensus workshop: copper. Eur J Clin Nutr 56(6):469–483

Schuytema GS, Nebeker AV, Griffis WL (1994) Effects of dietary exposure to forest pesticides on the brown garden snail *Helix aspersa* Müller. Arch Environ Contam Toxicol 26(1):23–28

Sheehan D (2006) Detection of redox-based modification in two dimensional electrophoresis proteomic separations. Biochem Biophys Res Commun 349:455–462

Shrivastava AK (2009) A review on copper pollution and its removal from water bodies by pollution control technologies. Indian J Environ Protect 29(6):552–560

Siddiqui NI, Yigzaw Y, Preaux G, Gielens C (2009) Involvement of glycans in the immunological cross-reaction between a-macroglobulin and hemocyanin of the gastropod *Helix pomatia*. Biochimie 91(40):508–516

Sigel A, Sigel H, Sigel RKO (2009) Role of metallothionein in metal metabolism and toxicology. In: Sigel A, Sigel H, Sigel RKO (eds) Metallothioneins and related chelators. Royal Society of Chemistry, Cambridge, pp 11–16

Simkiss K (1988) Molluscan skin (excluding Cephalopods). In: Wilbur K, Trueman E, Clarke M (eds) The Mollusca, vol 11. Academic Press Inc, New York, pp 11–34

Simkiss K (1985) Prokaryote-eukaryote interactions in trace element metabolism. *Desulfovibrio* sp. in *Helix aspersa*. Experientia 41:1195–1197

Simkiss K, Mason AZ (1983) Metal ions: metabolic and toxic effects. In: Hochanka PW (ed) The mollusca, vol 2, Environmental biochemistry and physiology. Academic, New York, pp 102–164

Simkiss K, Jenkins KGA, Mc Lellan J, Wheeler E (1982) Methods of metal incorporation into intracellular granules. Experientia 38:333–335

Simkiss K (1981) Calcium, pyrophosphate and cellular pollution. Trends Biochem Sci 6:3–5

Singh KP, Mohan D, Sinha S, Dalwani R (2004) Impact assessment of treated/untreated wastewater toxicants discharged by sewage treatment plants on health, agricultural, and environmental quality in the wastewater disposal area. Chemosphere 55(2):227–255

Singh RP, Kumar S, Nada R, Prasad R (2006) Evaluation of copper toxicity in isolated human peripheral blood mononuclear cells and it's attenuation by zinc: *ex vivo*. Mol Cell Biochem 282(1–2):13–21

Snyman RG, Reinecke SA, Reinecke AJ (2009) Quantitative changes in digestive gland cells and oocytes of *Helix aspersa*, as biomarkers of copper oxychloride exposure under field conditions. Bull Environ Contam Toxicol 83(1):19–22

Snyman RG (2001) Copper oxychloride, in the common garden snail *Helix aspersa*, in Western Cape Vineyards. PhD thesis, University of Stellenbosch

Snyman RG, Reinecke SA, Reinecke AJ (2000) Hemocytic lysosome response in the snail *Helix aspersa* after exposure to the fungicide copper oxychloride. Arch Environ Contam Toxicol 39(4):480–485

Soto M, Quincoces I, Marigómez I (1998) Autometallographical procedure for the localization of metal traces in molluscan tissues by light microscopy. J Histotechnol 21(2):123–127

Stehlik-Tomas V, Gulan Zetić V, Stanzer D, Grba S, Vahčić N (2004) Zn, Cu and Mn enrichment in *S. cerevisiae*. Food Technol Biotechnol 42(2):115–120

Strandberg B, Axelsen JA, Pedersen MB, Jensen J, Attrill MJ (2009) Effect of a copper gradient on plant community structure. Environ Toxicol Chem 25(3):743–753

Sturm CF, Pearce TA, Valdés Á (2006) Chapter 1: the mollusks: introductory comments. In: Sturm CF, Pearce TA, Valdés Á (eds) The Mollusks: a guide to their study, collection, and preservation. American Malacological Society. Universal Publishers, Boca Raton, Florida, USA, pp 1–7

Svendsen C, Spurgeon DJ, Hankard PK, Weeks JM (2004) A review of lysosomal membrane stability measured by neutral red retention: is it a workable earthworm biomarker? Ecotoxicol Environ Saf 57(1):20–29

Svendsen C, Weeks JM (1997a) Relevance and applicability of a simple earthworm biomarker of copper exposure: I. Links to ecological effects in a laboratory study with *Eisenia andrei*. Ecotoxicol Environ Saf 36(1):72–79

Svendsen C, Weeks JM (1997b) Relevance and applicability of a simple earthworm biomarker of copper exposure: II. Validation and applicability under field conditions in a mesocosm experiment with Lumbricus rubellus. Ecotoxicol Environ Saf 36(1):80–88

Svendsen C, Weeks JM (1995) The use of lysosome assay for the rapid assessment of cellular stress from copper to the freshwater snail *Viviparus contectus*. Mar Pollut Bull 31(1–3):139–142

Swaileh KM, Rabay'a N, Salim R, Ezzughayyar A, Rabbo AA (2001) Levels of trace metals and effect of body size on metal content of the landsnail *Levantina hierosylima* from the West Bank-Palestine. J Environ Sci Health A 36(7):1373–1388

Swaileh KM, Ezzughayyar A (2000) Effects of dietary Cd and Cu on feeding and growth rates of the landsnail *Helix engaddensis*. Ecotoxicol Environ Saf 47(3):253–260

Symondson WOC, Lidell JE (1993) The detection of predation by *Abax parallelepipedus* and *Pterostichus madidus* (Coleoptera: Carabidae) on Mollusca using a quantitative ELISA. Bull Entomol Res 83(4):68–78

Tam C, Idone V, Devlin C, Fernandes MC, Flannery A, He X, Edward Schuchman E, Tabas I, Andrews NW (2010) Exocytosis of acid sphingomyelinase by wounded cells promotes endocytosis and plasma membrane repair. J Cell Biol 189(6):1027–1038

Taylor MG, Simkiss K, Greaves GN, Harries J (1988) Corrosion of intracellular granules and cell death. Proc R Soc London B 234(1277):463–476

Teofilova T, Kodhabashev N, Gerasimov S, Markova E (2011) Comparative characterization of the heavy metal contents from samples in two regions in Bulgaria with different anthropogenic load. Natura Montenegrina 9(3):897–912

Titova I (2011) Vodokanal employs snails to monitor air quality. The St Petersburg Times 1642(4): Wednesday, February 9, 2011

Triebskorn R, Köhler HR (1996) The impact of heavy metals on the grey garden slug, *Deroceras reticulatum* (Müller): metal storage, cellular effects and semi-quantitative evaluation of metal toxicity. Environ Pollut 93(3):327–343

UniProtKB (2011) http://www.uniprot.org/uniprot/P55947. Accessed 4 July 2011

UNEP/RAMOGE (1999) Manual on the biomarkers recommended for the MED POL biomonitoring programme.UNEP. Athens: 1–92

Uriu-Adams JY, Keen CL (2005) Copper, oxidative stress, and human health. Mol Aspects Med 26(4–5):268–298

Valko M, Morris H, Cronin MT (2005) Metals, toxicity and oxidative stress. Curr Med Chem 12(10):1161–1208

Velkova L, Dimitrov I, Schwarz H, Stevanovic S, Voelter W, Salvato B, Dolashka-Angelova P (2010) Structure of hemocyanin from garden snail *Helix lucorum*. Comp Biochem Physiol A Mol Integr Physiol 157(1):16–25

Veselý J (2000) The history of metal pollution recorded in the sediments of Bohemian Forest lakes: since the Bronze Age to the present. Silva Gabreta 4:147–166

Viard B, Pihan F, Promeyrat S, Pihan J-C (2004) Integrated assessment of heavy metal (Pb, Zn, Cd) highway pollution: bioaccumulation in soil, *Graminaceae* and land snails. Chemosphere 55(10):1349–1359

Vidal PL, Divisia-Blohorn B, Bidan G (1999) Conjugated polyrotaxanes incorporating mono- or divalent copper complexes. Inorg Chem 38(19):4203–4210

Vinitketkumnuen U, Kalayanamitra K, Chewonarin T, Kamens R (2002) Particulate matter, PM10 and PM2.5 levels, and airborne mutagenicity in Chiang Mai, Thailand. Mutat Res 519(12):121–131

Walker CH, Hopkin SP, Sibly RM, Peakall DB (1996) Biomarkers. In: Walker CH, Hopkin SP, Sibly RM, Peakall DB (eds) Principles of ecotoxicology. Taylor and Francis, London, pp 175–194

Wang Q, Zhou D, Cang L, Li L, Zhu H (2009) Indication of soil heavy metal pollution with earthworms and soil microbial biomass carbon in the vicinity of an abandoned copper mine in Eastern Nanjing, China. Eur J Soil Biol 45(3):229–234

Warnken J, Dunn RJK, Teasdale PR (2004) Investigation of recreastional boats as a source of copper at anchorage sites using time-integrated diffusive gradients in thin film and sediment measurements. Mar Pollut Bull 49(9–10):833–843

Wegwu MO, Wigwe IA (2006) Trace metal contamination of the African giant land snail *Archachatina marginata* from Southern Nigeria. J Chem Biodivers 3:88–93

Weser U, Schubofz LM, Younes M (1979) Chemistry of copper proteins and enzymes. In: Nriagu JO (ed) Copper in the environment II. Health effects. Willey, New York, pp 197–232

Williamson P (1980) Variables affecting body burdens of lead, zinc and cadmium in a road side population of the snail *Cepaea hortensis* Muller. Oecologia 44:213–220

Wolda H, Zweep A, Schuitema KA (1971) The role of food in the dynamics of populations of the landsnail Cepaea nemoralis. Oecologia 7:361–381

WHO (2004) Copper in drinking-water. Background document for development of WHO guidelines for drinking-water quality (WHO/SDE/WSH/03.04/88)

Yetimoglu EK, Urucu OA, Yurtman GZ, Filik H (2010) Selective determination of copper in water samples by atomic absorption spectrometry after cloud point extraction. Anal Lett 43(12):1846–1856

Yildiz D, Kula I, Ay G, Baslar S, Dogan Y (2010) Determination of trace elements in the plants of Mt. Bozdag, Izmir, Turkey. Arch Biol Sci 62(3):731–738

Yoo JI, Kim KH, Jang HN, Seo YC, Seok KS, Hong JH, Jang M (2002) Emission characteristics of particulate matter and heavy metals from small incinerators and boilers. Atmos Environ 36(32):5057–5066

Zanette J, Goldstone JV, Bainy ACD, Stegeman JJ (2010) Identification of CYP genes in *Mytilus* (mussel) and *Crassostrea* (oyster) species: first approach to the full complement of cytochrome P450 genes in bivalves. Marin Environ Res 69 Suppl: S1–S3

Zhang XY, Tang XL, Zhao CL, Zhang G, Hu HS, Wu HD, Hu B, Mo LP, Huang L, Wei JG (2008) Health risk evaluation for the inhabitants of a typical mining town in a mountain area, South China. Ann N Y Acad Sci 1140:263–267

Zhang WD, Wang XF, Li Q, Jiang Y, Liang WJ (2007) Soil nematode responses to heavy metal stress. Helminthologia 44(2):87–91

Zhang SZ, Lipsky MM, Trump BF, Hsu IC (1990) Neutral red (NR) assay for cell viability and xenobiotic-induced cytotoxicity in primary cultures of human and rat hepatocytes. Cell Biol Toxicol 6(2):219–234

Zödl B, Wittmann KJ (2003) Effects of sampling, preparation and defecation on metal concentrations in selected invertebrates at urban sites. Chemosphere 52(7):1095–1103

Zvereva EL, Kozlov MV (2010) Responses of terrestrial arthropods to air pollution: a meta-analysis. Environ Sci Pollut Res Int 17(2):297–311

Index

Printed in the United States
By Bookmasters